Digital Agriculture, Methods and Applications

Edited by Redmond R. Shamshiri and Sanaz Shafian

Published in London, United Kingdom

Digital Agriculture, Methods and Applications
http://dx.doi.org/10.5772/intechopen.98141
Edited by Redmond R. Shamshiri and Sanaz Shafian

Contributors
Seyed Moin-eddin Rezvani, Redmond R. Shamshiri, Ibrahim A. A. Hameed, Jalal Javadi Moghaddam, Siva K. Balasundram, Galina Yakubova, Aleksandr Kavetskiy, Nikolay Sargsyan, Stephen A. Prior, Henry Allen Torbert, Wipa Tangkananond, Thanat Na Phatthalung, Mohammed El Idrissi, Omar El Beqqali, Jamal Riffi, Sanaz Shafian, Denis Kotarski, Petar Piljek, Marko Pranjić, Tomislav Petanjek

First published in London, United Kingdom, 2022 by IntechOpen
IntechOpen is the global imprint of INTECHOPEN LIMITED, registered in England and Wales, registration number: 11086078, 5 Princes Gate Court, London, SW7 2QJ, United Kingdom

British Library Cataloguing-in-Publication Data
A catalogue record for this book is available from the British Library

Additional hard and PDF copies can be obtained from orders@intechopen.com

Digital Agriculture, Methods and Applications
Edited by Redmond R. Shamshiri and Sanaz Shafian
p. cm.
Print ISBN 978-1-80355-462-4
Online ISBN 978-1-80355-463-1
eBook (PDF) ISBN 978-1-80355-464-8

Meet the editors

Dr. Redmond R. Shamshiri is a scientist at the Leibniz-Institut für Agrartechnik und Bioökonomie working toward digitization of agriculture for food security. He holds a Ph.D. in agricultural automation with a focus on control systems and dynamics. His research fields include distributed collision avoidance for autonomous navigation of agricultural mobile robots, LPWAN sensors for digital farming applications, and optimization of controlled environment crop production systems. He is currently working on the adaptation of the 5G technology for reducing yield losses from potato fields.

Dr. Sanaz Shafian is an Assistant Professor of Smart Farming at Virginia Tech University. Prior to this, she was an assistant professor at the University of Idaho. Her expertise lies in remote sensing research, with a focus on using advanced technologies and methodologies for economically and environmentally sustainable crops and tree monitoring and management. She integrates satellite/drone images and AI to develop methodologies for environmental monitoring, crop modeling, and water and nutrient conservation and she has published widely on these topics. She has been involved in several USDA projects. At the University of Idaho, she led an educational and outreach project to initiate a Precision Agriculture certificate. Sanaz is currently working on two agriculture-related projects, one mapping invasive plants in temperate forests and evaluating forest health in Virginia and another looking at developing smart tools for the early detection of soybean diseases.

Contents

Preface

Digital Agriculture (DA) refers to the practice of modern technologies such as sensors, robotics, and data analysis for improving the sustainability and profitability of farms, while at the same time increasing crops' yield and quality. Digitalization of agriculture is a technological response to climate change, global warming, and water scarcity that are affecting agricultural productivity and food security. Reports indicate that nearly one-quarter of the global greenhouse gas emissions comes from crop cultivation and livestock farming, which can significantly decrease crop yields. In modern farms, measurements from multiple in-situ sensors combined with images, maps, and data generated by satellites, drones, artificial intelligence, and prediction models are delivering detailed agronomic information on crop conditions and field variabilities to improve various aspects of farming practices whilst diminishing risks and uncertainties. The inputs and outputs of digital agriculture, as shown in **Figure 1**, have evolved based on data streams and flexible data-sharing services and contributed to mitigation strategies for climate change by providing a series of scientific solutions toward reducing pesticide usage, chemical fertilizers, and minimizing energy demands.

Until 2010, growers had to rely on Global Positioning System (GPS), ground-based sensing platforms, satellite maps, and local sensing devices such as data loggers to monitor their fields, identify deficiencies, and improve crop yield via better management of the resources (i.e., variable rate technology). These practices were referred to as Precision Agriculture (PA) and Smart Farming. With the rise of Unmanned Aerial

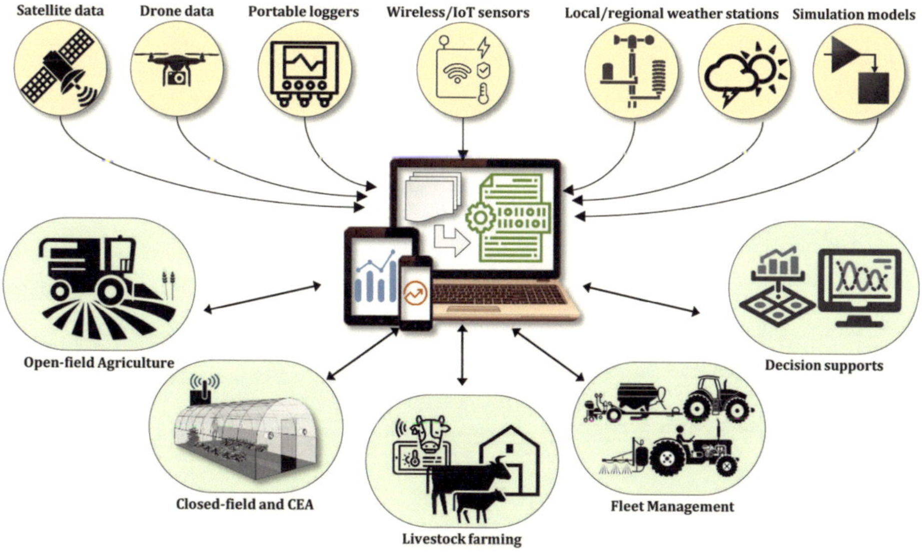

Figure 1.
Schematic demonstration of the inputs and outputs of digital agriculture from a general perspective (Source: AdaptiveAgroTech).

Vehicles (UAV), low-powered long-range wireless sensors, IoT gadgets, and advances in robotics, PA concepts and methods shifted toward digitization and contributed more to the economic development and sustainability of food production. By 2012, digital agriculture began to incorporate a wider variety of technological advances such as small-scale robots, swarm drone technology, distributed wireless networks, cloud-based automation, and mobile apps in order to continuously monitor, evaluate, and manage soil condition, water resources, and weather fluctuations on the farmlands to enhance field productivity and reduce operational costs. More recently, digital agriculture developed a series of methods based on artificial intelligence and machine learning to analyze and interpret high-resolution drone NDVI imagery and data (**Figure 2**) for monitoring crop water level and quality, determining soil moisture and soil salinity, creating yield maps, health assessment, and crop stress identification. On the automation side, wireless sensors and IoT devices have been used for smart irrigation, water loss management, and continuous identification of soil nutrient contents in remote areas.

With the introduction of the fifth-generation mobile network (5G), digital agriculture is redefining some of the concepts of the sense-think-act paradigm in the fields. One of the trending topics in this context is the deployment of distributed automation systems such as collaborative robots and a swarm of small-scale unmanned machinery that can autonomously execute various site-specific operations such as weeding and spraying via IoT-based cloud computing services. While similar solutions are being implemented as pilot plant projects or on commercial scales, connection stability and security between nodes have been always a concern. A review of the literature reveals that the use of robots in agriculture with modular electronic control units is growing

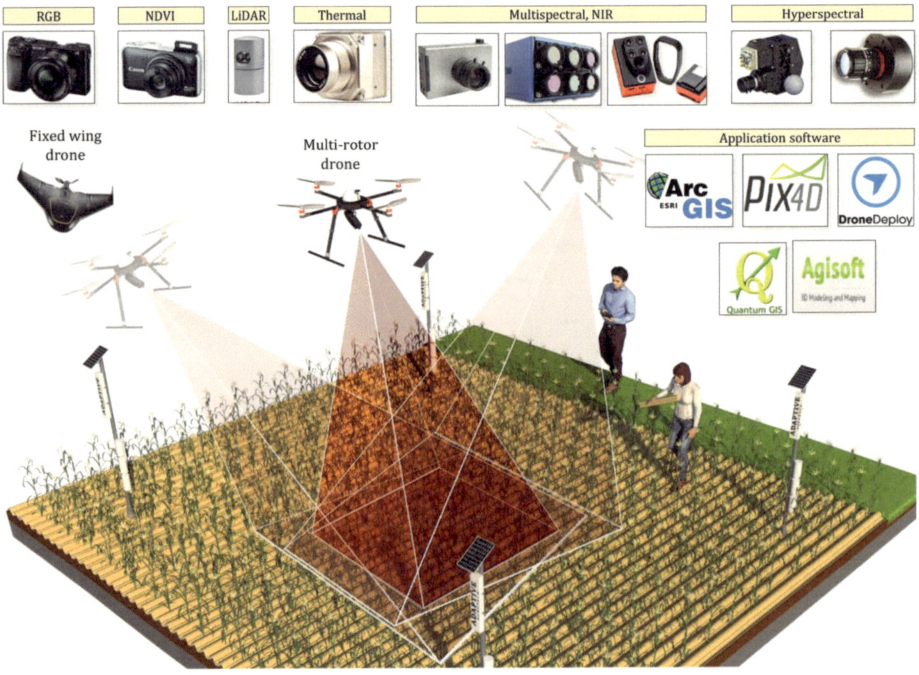

Figure 2.
Illustration of UAV-based photogrammetry for estimation of crop parameters via nadir and oblique views (Source: AdaptiveAgroTech).

rapidly and becoming an active field of research, drawing design attention to affordable components that can be easily replaced upon failure. These robots are expected to identify deficiencies and variations in large-scale cultivations and to respond to them with precision technology and site-specific management solutions. For this purpose, autonomous mobile robots that are equipped with various data acquisition devices, multi-spectral cameras, and Light Detection and Ranging (LiDAR) sensors provide a great opportunity for field scouting, health assessment, early disease detection, and yield estimation. In addition, these robots can be integrated with custom-built end-effectors and manipulators to perform specific tasks such as mowing, weeding, and spraying. In addition, mobile robots for digital agriculture are required to withstand harsh field conditions, have a flexible control design with interchangeable and compatible components, and benefit from a reliable navigation system with collision avoidance capabilities. In addition, farmers prefer that depending on the task requirements, different modules such as sensors, actuating devices, and manipulators can be easily swapped on a multi-purpose robot.

Digital agriculture is offering significant potential to replace conventional farming methods with cutting-edge technologies toward creating farms of the future that are expected to be connected and be zero CO2 emissions. If successfully integrated and implemented, digital agriculture can also play a key role in reducing agricultural production costs by decreasing the number of human workforces that are currently engaged in performing repetitive tasks. The presented book aims to expand and highlight these aspects from an academic perspective in separate chapters. Most of the solutions and strategies described in this book represent a valuable aspect of digital agriculture that is aiming at preserving natural resources and securing food production for the increasing world population.

Redmond R. Shamshiri
Technische Universität Berlin,
Berlin, Germany

Sanaz Shafian
Virginia Tech,
Virginia, United States of America

Section 1

Perspective

Chapter 1

Digital Agriculture and Intelligent Farming Business Using Information and Communication Technology: A Survey

Mohammed El Idrissi, Omar El Beqqali, Jamal Riffi, Redmond R. Shamshiri, Sanaz Shafian and Ibrahim A. Hameed

Abstract

Adopting new information and communication technology (ICT) as a solution to achieve food security becomes more urgent than before, particularly with the demographical explosion. In this survey, we analyze the literature in the last decade to examine the existing fog/edge computing architectures adapted for the smart farming domain and identify the most relevant challenges resulting from the integration of IoT and fog/edge computing platforms. On the other hand, we describe the status of Blockchain usage in intelligent farming as well as the most challenges this promising topic is facing. The relevant recommendations and researches needed in Blockchain topic to enhance intelligent farming sustainability are also highlighted. It is found through the examination that the adoption of ICT in the various farming processes helps to increase productivity with low efforts and costs. Several challenges are faced when implementing such solutions, they are mainly related to the technological development, energy consumption, and the complexity of the environments where the solutions are implemented. Despite these constraints, it is certain that shortly several farming businesses will heavily invest to introduce more intelligence into their management methods. Furthermore, the use of sophisticated deep learning and Blockchain algorithms may contribute to the resolution of many recent farming issues.

Keywords: intelligent farming, food security, fog/edge computing, blockchain, digital twin, artificial intelligence

1. Introduction

Recently, the agricultural domain is facing numerous challenges related to the need to permanently increase productivity, climate change management, crop health monitoring, and irrigation water management, as well as fertilization optimization. To address these constraints, IoT technology is opening up new promising technological paths and pushing the future of agriculture to the next level. Indeed, many advantages are offered by IoT systems for intelligent farming, such as a panoply of sensor networks to optimize irrigation and agricultural inputs management, as well as improvement of the agricultural engine guidance and maintenance. Agricultural sensors implemented in the fields are estimated to reach 12 million by 2023, this revolution of smart devices will provide many remote facilities to manage seeds, irrigation, fertilizers, and early disease detection by collecting real-time data about the field and the environment. We mean by intelligent farming the integration of smartness in the farming processes, not only for the land management but also in the other chain links notably logistics and supply chain, transportation [1–4], as well as storage. The need for automation in the agricultural domain to overcome the constraints imposed by classical methods of farming became more essential than before. Furthermore, the availability of water in a sufficient quantity and quality has been recently become alarming because of the climate change phenomenon. Consequently, many technological, economical, and social policies have to be implemented according to many recent studies that focused on water management topics [5]. Thanks to the smart and low-cost dedicated sensors, irrigation tasks will be precise and the productivity will be rapidly increased, without ignoring the important contribution in hydrological resources preservation. Traceability of the food supply chain (FSC) is an important key factor to ensure the quality and safety of food transportation and identification in a regulatory manner, as well as protect perishable food against waste. Dairy farming is another farming process that has taken benefits from the integration of information and communication technology in the farming industry, it helps farmers to adopt more accurate practices in dairy management [6] to monitor the heat of oestrus to improve reproduction, as well as the animal health check and monitoring [7].

Highly intelligent farming or high intelligent farming are two concepts that refer to the use of high tech in farming processes to enhance the efficiency of daily work. In fact, using technological innovation in farming is not new, but the rise of some disciplines, such as IoT, fog computing, satellites, drones, smartphones, and Blockchain, are things that will push smart agriculture and precise farming industries to a high level in the coming years. We believe that implementing ICT in the farming world will enable farmers to better understand and interact with their farms by collecting data about changing variables and giving commands according to the situations. All of these technologies will give the ability to the farms to make a big transition from being simple physical environments to highly intelligent and abstract worlds.

Despite the existence of several studies and surveys that introduce the issue of integrating ICT in farming processes, we find that these surveys either focused only on one farming process or do not investigate deep enough this integration. Moreover, investigation of Blockchain technology, its benefits for the farming industry, and its required research to build sustainable development, need to be elaborated. To fill these literature gaps, we propose this survey as one of the most mature studies of its kind that presents a systematic and developed state-of-the-art for integrating ICT in the farming world.

The remainder of this work is further structured as follows: The research methodology is presented in Section 2. Then, the general review of IoT-based systems' requirements is discussed in Section 3. Next, Section 4 provides the components of an intelligent farming IoT model. After that, the open challenges resulting from IoT-fog computing integration are discussed in Section 5. The applications of Blockchain in intelligent farming and the discussion part are then presented in Section 6. The conclusion and summary are provided in Section 7.

2. Research methodology

This survey extensively studies the knowledge related to the intelligent farming domain. It inventories and summarizes the integration of ICT in the IF field. The potential of this survey regarding the other works is to evaluate the implementation of Blockchain in the IF topic.

2.1 Reference management

The references related to our research area are collected and filtered, 104 references have been retained based on the following four criteria: (1) High priority was given to recent studies, which means that most of the selected papers were published between 2017 and 2021, and some of them are in press. (2) The timeliness and novelty of the study in the intelligent farming field is another criterion that has been given more priority. (3) The significance to the field and the potential impact on the course of future work in the area of smart farming, were also criteria that have been taken into account while selecting the examined papers. (4) Since the potential of our survey is the evaluation of the applications and benefits of Blockchain technology for the farming industry, we have given more importance to the studies that have explored Blockchain technology within farming environments. A variety of questions that are addressed in this survey can be summarized as follows—(Q1) what type of ICT systems and frameworks are used in the implementation of IF solutions? The answer to this question gives a general study of relevant technologies and protocols adopted in IoT systems as well as fog/edge computing platforms. These technologies represent the basis of many implementations in many fields including intelligent farming, (Q2) is there an IoT model structure that can be adopted to build IF solutions? To answer this question, a five-layer model for intelligent farming is presented, (Q3) how Blockchain technology can be used in the IF domain, and what impact might this have on IF practices improvement? The answer to this question leads us to introduce the

Source	Hardware and protocols	Cloud computing	Fog computing	Blockchain
Ratnaparkhi et al. [8]	✓			
Tahsien et al. [9]	✓			
Hajjaji et al. [10]	✓	✓		
Farooq et al. [11]	✓	✓	✓	
Mekala et al. [12]	✓	✓		
Cisternas et al. [13]				
Lova Raju et al. [14]	✓	✓		
Shi et al. [15]	✓	✓	✓	
Muangprathub et al. [16]	✓	✓		
Bacco et al. [17]		✓	✓	
This survey	✓	✓	✓	✓

Table 1.
Comparison between this survey and other previous surveys.

most recent novelty of Blockchain usage in the IF domain, as well as the challenges and the needed researches to enrich this debate.

2.2 Comparison with other smart agriculture state-of-the-arts and reviews

Starting from the examined papers, we have identified several state-of-the-arts, surveys, and reviews, each type of those papers discussed the use of ICT in intelligent farming based on specific ICT disciplines. Some previous surveys focused on the hardware used to implement IF applications, and others covered the integration of IoT with fog/edge technologies to optimize some metrics. Some points are common between our work and others, such as the description of the hardware and protocols adopted in IF systems, and the implementations of IF applications in cloud/fog computing environments. In this work we studied the Blockchain discipline related to the farming domain, this point has not been obviously covered by the other surveys. **Table 1** summarizes the comparison between this survey and the other previous works.

3. A general review of IoT-based systems' requirements

In most cases, precision agriculture data are communicated wirelessly between sensors, or between IoT devices and the core using several kinds of communication protocols, these protocols define the rules and the different formats of the communicated data. The secret behind the success of IoT systems is the development of communication protocols [18, 19], such as RFID (Radio Frequency Identification), NFC (Near Field Communication), IEEE 802.11 Wi-Fi, IEEE 802.16 Wi-Max (Worldwide Interoperability for Microwave Access), IEEE 802.15.4 LR-WPAN (Low-Rate Wireless Personal Area Networks), 4G and 5G cellular networks, IEEE 802.15.1 Bluetooth, ZigBee, ANT/ANT+ networks, DASH7, Enocea...).

Unlike the fog computing paradigm, the traditional cloud computing approach is characterized by centralization, high latency, and more network failures. These characteristics among others make cloud computing unsuitable for IoT applications where time and mobility are crucial factors. In the IoT context, fog computing is a new computing approach that helps to distribute the load of processing and make it so close to the sensing layer. One of the solutions that were proposed to accelerate the processing and compensate for the resource limitation of IoT devices is computation offloading. This concept allows devices to fully/partially offload their computation tasks to resource-rich cloud infrastructures [20]. But this solution bypasses only the cloud computing limitations and does not propose a real solution to resolve them. A group of researchers [21] discussed the usefulness of another concept called computation onloading. This concept is based on bringing cloud services to the edge of the network to satisfy the requirements of IoT devices in terms of bandwidth and latency.

Many contributions are proposed to improve the shared characteristics between cloud and fog computing, notably the generated latency between requesting the task execution and receiving the response, the energy consumed during the task processing, the resource management strategy that defines the provided quality of service, the security issue directly linked to the privacy of generated data, the mobility support to ensure the best quality of service to the end devices, the interoperability between smart things, the scalability related to the exponential increase of the number of IoT devices, and finally the bandwidth needed to transmit data from the network of smart objects to the processing center.

The latency generated by the cloud is significantly important, this is an issue for new IoT mobile applications that need real-time responses to their requests. To enhance this characteristic through the fog/edge computing model, Yang et al. [22] developed an offline heuristic algorithm, SearchAdjust, to minimize the average latency for Multiuser Computation Partition Problem (MCPP). In the same context, Yousefpour et al. [23] developed and evaluated a policy to reduce the service delay for IoT devices based on offloading and sharing load approach. In another work, Molina et al. [24] proposed a strategy of uplink/downlink, and edge computational resources allocation in a multi-user scenario to achieve latency and energy efficiency in task processing. Ren et al. [25] investigated the collaboration between cloud computing and edge computing, where the tasks of mobile devices can be partially processed at the edge node and the cloud server. A joint communication and computation resource allocation problem is formulated to minimize the weighted-sum latency of all mobile devices.

Regarding the energy consumption issue, most of the processing tasks are carried out in the cloud computing data centers that increase the quantity of energy needed for query transmission and execution. This consumption is minimized in the fog/edge computing model because the majority of computing tasks are distributed over several end devices or offloaded to the edge mini data centers. In this context, Xiang et al. [26] proposed a policy to efficiently optimize energy in LTE (Evolution Long Term)/Wi-Fi link selection and transmission scheduling, as well as developed an approximate dynamic programming algorithm to reduce energy consumption in the MCC (Mobile Cloud Computing). Ge et al. [27] proposed a game-theoretic strategy to reduce the overall energy dissipation of both mobile devices and cloud servers considering the offloading technique in the MCC system. Chen et al. [28] adopted a game-theoretic approach to propose a multi-user offloading solution for mobile-edge cloud computing, their proposed solution aims to achieve energy efficiency in a multi-channel wireless interference environment.

In the classical cloud computing approach, the efficiency of resources management is less compared to the fog/edge computing approach, this is due to the existence of more sophisticated algorithms that proved their efficiency in resources allocation. In this window, Mostafa et al. [29] proposed an automated fog selection and allocation scheme of task requests by IoT devices. In another work, Jana et al. [30] proposed a QoS (Quality of Service)—aware resource management technique for the efficient management of resources. Souza et al. [31] developed a scheme that combines fog computing and cloud resource allocation. Aazam et al. [32] proposed a user characteristic-based resource management for fog, which performs efficient and fair management of resources for IoT deployments. Delegating data protection to the cloud layer without implementing mechanisms to protect data at the end device level is an inefficient strategy. The best way is to ensure end-to-end data protection, the fog/edge computing model is mainly concerned by this issue compared to the cloud computing approach that focuses on data protection at the cloud level. Das Manik [33] proposed a security protocol for IoT applications based on Elliptic Curve Cryptography (ECC). Hernández-Ramos et al. [34] proposed a new mechanism of lightweight authentication and authorization to be embedded in a smart object based on DCapBAC (Distributed Capability-Based Access Control). Zhang et al. [35] suggested using Ciphertext-policy attribute-based encryption (CP-ABE), which is a recognized cryptographic technique to ensure data confidentiality and provide firm access control.

The majority of IoT devices used in smart cities or smart environments are geographically distributed, mobility of IoT devices and applications should, therefore, be supported by the adopted computing approach. As a result, many works are proposed to enhance the mobility of end devices in the fog/edge model since this characteristic is less present in the traditional cloud computing model. For this purpose, Chaisiri et al. [36] proposed a mobility-aware offloading priority design, it aims to precisely anticipate users' mobility profiles and channels. In the same context, Prasad et al. [37] proposed an approach for mobility management along with traffic control to offer better users' QoE (Quality of Experience) with latency-tolerant tasks. Ning et al. [38] constructed a three-layer VFC (Vehicle Fog Computing) model to enable distributed traffic management and minimize the response time of citywide events collected and reported by vehicles.

Interoperability is another important difference between the fog/edge computing model and the cloud computing approach regarding provided smart services. The interoperability requires that all interfaces of cloud-based or fog/edge-based systems are wholly understood. Despite that cloud computing offers more interoperability for some distributed applications, it is difficult to cover smart things applications due to the big heterogeneity of manufacturers and systems. Contrary to cloud computing, fog/edge computing is more open to the end devices and tends to ameliorate the interoperability issue in an IoT system. Starting from this requirement, Jayaraman et al. [39] proposed an OpenIoT platform used for the digital agriculture use case (Phenonet), the OpenIoT enables semantic interoperability. Desai et al. [40] proposed a semantic web permit architecture to afford interoperability among smart things. Ullah et al. [41] proposed a semantic interoperability model for big-data in IoT (SIMB-IoT) to deliver semantic interoperability among heterogeneous IoT devices in the health care domain.

In the traditional cloud computing model, the number of smart supported devices and applications increases at a slow rate oppositely to what happens in fog/edge computing systems. Scalability is an essential feature that defines how resources provisioning is performed and what components can be scaled, notably the storage capacity, the number of fog/edge nodes, the connectivity solutions, and the internal hardware or software of fog/edge nodes. Tseng and Lin [42] designed a mechanism to dynamically scale in/out the serving instances of the middle nodes to make the whole IoT/ M2M (Machine to Machine) platform more scalable using an industrial IoT (IIoT) scenario. Vilalta et al. [43] proposed a new fog computing infrastructure named TelcoFog that can be installed at the edge of the mobile network of the telecom operator to provide several services, such as NFV (Network Function Virtualization) and MEC for IoT applications, the benefits of the proposed infrastructure are dynamic deployment, scalability, and low latency. Gupta et al. [44] proposed a highly distributed service-oriented middleware called SDFog (Software-Defined Fog) based on cloud and fog capabilities as well as SDN (Software-Defined Networking) and NFV to satisfy the required high level of scalability and QoS.

The bandwidth needed to transmit the data collected is closely tied to the generated latency, the biggest amount of data requires more bandwidth to be transmitted to the cloud data centers, which means more latency in the transmission process. Optimizing the bandwidth in a fog/edge environment directly minimizes the delay resulting from the transmission process because the processing resources are located close to the end devices. In this context, Ito et al. [45] proposed a bandwidth allocation scheme based on collectible information. Gia et al. [46] introduced the processing of ECG (electrocardiogram) features using fog nodes, their results disclosed that fog helps to achieve more efficiency in bandwidth and low latency in the data processing. Bhardwaj et al. [21] argued the utility of "onloading" cloud services to the edge of the network to address the bandwidth and latency challenges of IoT networks.

4. Components of an intelligent farming IoT model

Before deciding to integrate IoT infrastructure in a given smart farming business model, it is first mandatory to understand the components of the IoT model, because this is the best way to analyze business technology compromises, and better define the requirements of the farming process system. **Figure 1** illustrates the five layers comprising of the smart farming IoT model, each layer is explained in greater detail below.

4.1 Hardware sensor and actuator layer

This component is located in the bottom layer of the IoT model, it can also be called the data collection and actuation layer, it is considered as the link between the farm physical world and virtual data management and decision making. Functionally, this layer is responsible for sensing capabilities to gather data about the physical farming variables that we want to measure, as well as take actions to change the environment depending on the scenario of the made decision. In this layer, it is recommended to take into account the hardware characteristics, such as size, cost, useful lifetime, reliability, performances, as well as the scenario of use. Physical sensors existed for a long time before even the emergence of IoT devices, the only difference is that their uses have become more sophisticated and they have been used more ubiquitously. The intelligent farming sensors can be manufactured separately or embedded in a specific

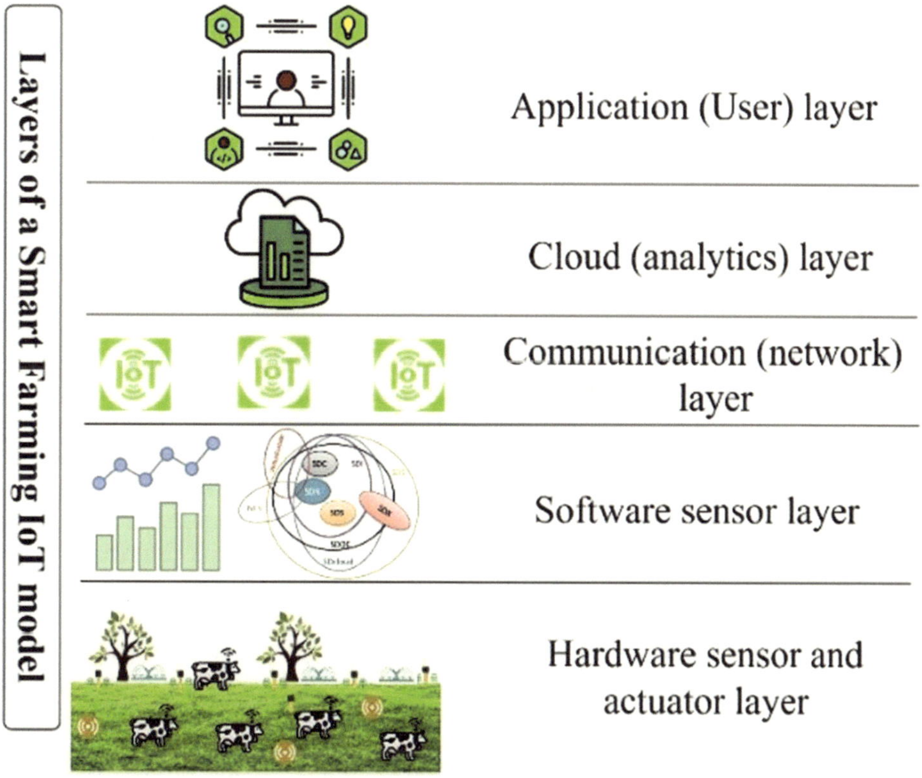

Figure 1.
The five layers of a smart farming IoT model.

one board and dedicated to a particular application. The common applications of sensors are to measure temperature, humidity, geographical position, light and sound sense, and much more.

The farming actuators are the translators of the decision to comprehensive and useful energy capable to change the environment from one condition to another, such as guiding an agricultural engine, changing the temperature, making a movement, or enabling/disabling a pump. Operationally speaking, actuators can take three forms—pneumatic using air pressure, electrical using electrical energy, and hydraulic based on the power of liquids.

4.2 Software sensor layer

This layer represents the point of connection between the physical world and the fog-cloud environment, it defines how an object can be smart by doing local analytics, take simple decisions, or control other devices. This layer enables the "software-defined hardware infrastructure (SDHI)" or "resource desegregation" [47] concept, which is one of the software-defined environment taxonomy. This concept is of great interest today because it considers physical hardware as a modular component offering more flexibility, agility, automation, and optimization in cloud resource allocation. It provides a new pool of resources-based vision and strategy to efficiently manage available hardware resources to serve multiple applications, this offers more programmability to the infrastructure. It exists in literature more similar concepts like virtualization technique [48, 49], Virtual Network Function (VNF) [50, 51], Software-defined cloud (SDCloud) [52, 53]. This layer is important and critical at the same time. Important because it can be used to minimize the hardware complexity, in other words, instead of being stuck in a fixed hardware architecture which is complex and expensive to build in most of the time, it is possible to design generic hardware like Field Programmable Gate Arrays (FPGA) and program it for various scenarios. And critical because it is the only gate through which the data flows from the physical world to cloud or fog environments, thus the definition of an OS (Operating System) that manages the hardware and the running applications is considered a critical task.

4.3 Communication (network) layer

In some contexts, this layer is called connectivity, it defines the manner of how data are sent and received between the cloud and the smart devices. The connectivity function has resulted from the combination of two essential elements—protocols and physical hardware used to transmit the signals. In the beginning, RFID is used by the objects to communicate with each other [54] without human intervention. With the emergence of 5G cellular network, a great opportunity is offered to accelerate the IoT systems' development, particularly with the emergence of the MTC (Machine Type Communication) concept, which is also called machine to machine communication, it refers to automated data communications among devices. According to the 3GPP (3rd Generation Partnership Project), it exists two modes of communications in MTC applications—the first mode can occur between an MTC device and a server, and the second can happen between a network of MTC devices [55]. Choosing the communication mode and protocol is a critical task for IoT project owners. This modeling step defines not only the communication with the cloud but also determines how IoT objects communicate with each other. Many communication technologies can be used, for instance, Bluetooth, ZigBee, Wi-Fi, and optical wireless communication for small coverage areas [56, 57]. Sigfox [58] and LoRa, LoRaWan (Long Range Wide Area Network) [59] have been conceived for a wide coverage area. Moreover, 5G is adopted to enhance all traditional mobile communication performances, and respond to multiple connectivity requirements of IoT applications, such as introducing low latency and reliability.

The heterogeneity in communication protocols as well as the complexity of manufacturers' models lead us to think about solutions to ensure the interoperability between IoT platforms and services. Consequently, the IoT middleware concept is immerged and many solutions have been proposed. The propositions can be classified into three big families [60]: Actor-based IoT middleware, cloud-based IoT middleware, and service-based IoT middleware. The first proposition of the actor-based middleware project offers an easy deployment in the distributed environments since it uses actor or agent concept, this middleware plays the role of a bridge between IoT devices and cloud services, it first works presciently to correctly receive data from each IoT device. It next sends the collected data to the cloud using HTTP (Hypertext Transfer Protocol) over TCP/IP protocol. The second family enables the terminology of the cloud of things (CoT) that was introduced by Yuriyama et al. [61], it is an enabler that lets us exploit and manage wireless sensors homogeneously without worrying about the manufacturer's physical complexities. CoT uses cloud capabilities in terms of elasticity of resource provisioning as well as automation, scalability, and cost-effectiveness. Considering this family of IoT middleware, the access of IoT devices to the cloud resources is ensured by the Application Programming Interface (API) of the cloud service provider or through the product vendor's application, as shown in **Figure 2(a)**.

The last family of IoT middlewares refers generally to the open-source platform named OpenIoT project, the objective behind proposing the SaaS (Sensing as-a-Service) solution is to find an adequate way to extract data from virtual cloud sensors without worrying about the physical architecture of the sensor that was behind the collected data. The architecture of the service-based IoT middleware is given in **Figure 2(b)**.

The most common criteria that is recommended to put in mind while choosing the adequate IoT middleware are stability regarding the application, the deployment mode (open source or commercial), the payment model (by the number of device/messages or using pay as you use mode), the level of security needed (depends on the criticality of the application and the managed data), the hardware compatibility

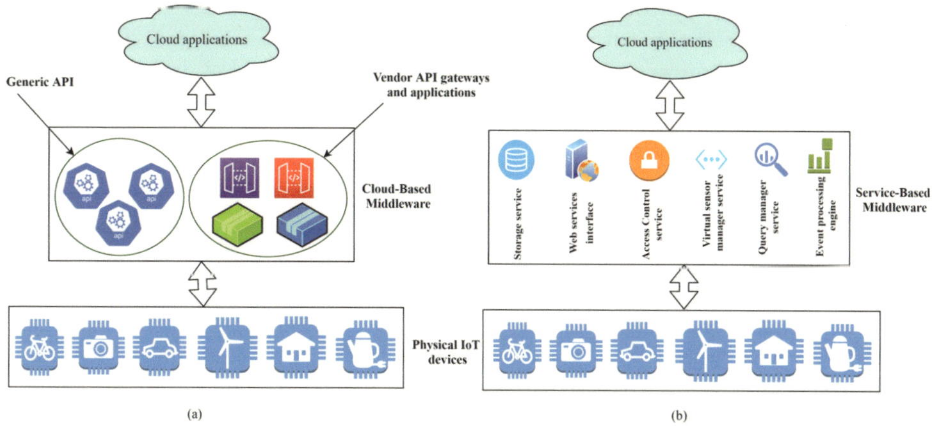

Figure 2.
Cloud-based and service-based IoT middleware.

(some commercial IoT Middlewares support the integration of some kind of hardware devices like Arduino and Raspberry), the protocol that the application requires (since it exists multiple types of communication protocols, some of them are open and others are proprietary), and either the middleware platform supports the required analytics or not (it depends on the nature of data that the application need which can be in real-time or historic).

4.4 Cloud (analytics) layer

IoT applications produce periodically what we call big data and send them to the backbone to be managed. The challenge for an IoT project manager is to consider many critical factors to conceive the right cloud architecture. This layer should take into account the essential 5 V of big data from the beginning, the 5 V as mentioned in Ref. [62] includes volume, variety, velocity, veracity, and value. The designed cloud architectures for IoT applications take many models depending on the project manager's perspective.

The model can be SaaS (Software as a Service), the customer in this case, does not have any knowledge about the platform architecture, the client only has a web interface or an API to interact with the provider platform, this model, in general, requires additional fees and the client still stuck in "Vendor lock-in," this means that more complexity and costs will be charged by the client if for any reason, decides to switch to another service provider.

The second model is PaaS (Platform as a Service), the client in this case has multiple choices of software bricks that can be used on-demand to build IoT applications without worrying about server management. This model provides many bricks for IoT solutions such as device management, storage, connectivity with other IoT fleets, collection, and transmission, as well as some machine learning options for decision-making support. The advantage of this model is the great ability offered by the vendor to the client to customize the IoT applications based on the offered software catalogs. But unfortunately, this can have some additional hidden costs.

The third kind of model is licensed or on-premise. Here, the vendor only makes support available to the client. The client buys software packages and the license, then installs them in his own managed infrastructure. All the maintenance tasks are under the client's responsibility. The open-source solutions are identical to the licensed model, the only difference is that the software packages are freely available, the solution maintenance is ensured by a community of volunteer developers. In some cases, the maintenance is performed by an enterprise and proposes the solution as a free package, while providing a paid version with other options. The tailor-made feature is another option adopted by many customers, it consists of engaging an external integrator to entirely conceive the IoT solution. In this case, the source code is owned by the application owner, he can use it subsequently to achieve the project evolutivity.

4.5 Application (user) layer

This layer is the most closer to the customer, it is generally used to ensure user-machine interaction, it defines how data is presented to the end-user depending on the user's requirements. In most cases, this layer is a web-based application. Some users require desktop, mobile, or wearable applications. Practically, the application layer is hosted somewhere in the provider's cloud to ensure the AAA

(Anytime, Anywhere, Application) capability. The most important thing that the IoT solution designer should understand is what the final users attend from the solution, and how this job can be done.

5. Open challenges of IoT-Fog integration in the IF context

5.1 Real-time processing

Fog computing provides required resources at the edge of the network to deliver real-time services for demanding applications (e.g, video streaming, gaming, video analytics, and robot-fog interactions [63]). When it comes to IoT data processing on a large scale, we can distinguish between three processing concepts [64], as illustrated in **Figure 3**.

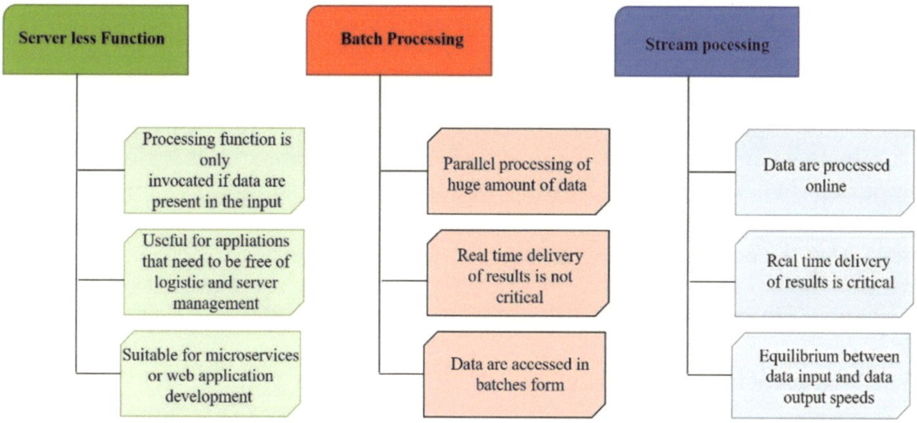

Figure 3.
Available modes of processing for IoT applications.

The serverless function also called Function as a service (FaaS), refers to the simplest processing model where data are present in the input of a black box, the results of processing are then gathered in the output without any session stat. The second processing concept is called batch processing, here, data are processed in small parts and often simultaneously, this type of processing is considered in situations when a large amount of data need to be processed, input data are accessed in batches form, or data need complex processing. The last processing mode is called stream processing, it refers to on-the-fly processing where data are processed online and the results are delivered instantly, this mode of processing is appropriate in case of real-time results are needed. Since IoT applications are diversified and data are generated and sent continuously to fog computing nodes, each processing mode can be adopted for a specific scenario.

5.2 Resource scheduling and management

It was expected that a huge number of IoT devices will be online shortly, meaning that the amount of generated data will be also colossal. Resource management policy is a determining factor in evaluating the quality of service delivered to IoT devices and applications. This policy depends on many factors such as the nature of the application requiring the resource. If the application allows delay of processing, all its requests are forwarded to the cloud resources to be executed there. But if the application is time-sensitive, all its requests are served by fog computing nodes.

5.3 IoT geo-distribution and mobility

Geo-distribution is one of the primary characteristics of smart devices. An object is most of the time moving from one geographical area to another, this mobility generates delay and packet loss [57]. Fog computing has to provide necessary mechanisms and resources to facilitate fog users' access at anytime, anywhere, and without any delay or loss, given that devices are highly distributed, handover is a critical mechanism among others that should be taken into account while conceiving and implementing fog computing architecture.

5.4 Latency minimizing

Most IoT devices have resource limitations in terms of communication, storage, and computation. As a direct result, the connected object needs a powerful infrastructure that can provide these requirements within a milliseconds scale. Cloud computing is known for its big latency, which makes it unsuitable for time-sensitive applications. On the other hand, the fog computing challenge is to provide necessary resources at the edge of the network to process data and serve IoT devices' requests within milliseconds to a few seconds scale. Fog computing serves also the central cloud by sending reports for data visualization purposes [65].

5.5 Security and integrity enhancement

Recently, IoT-generated data may represent the secret of an individual or an industry, indeed, they need to be protected in the transit phase and in-rest. The fog computing paradigm must ensure confidentiality-integrity and availability of data through efficient cryptographic algorithms. The security mechanisms offered by fog have to be light and less resource-consuming to be more adapted to the limited properties of end devices. In another hand, collected data are analyzed and treated locally in fog data centers instead of sending them through the internet to the cloud datacenter, this point helps a lot in data security reinforcement.

5.6 High availability

The exponential rise of IoT-generated data demands a reliable platform that can manage this huge amount of data. The temporary loss of connection is not an issue in the case of cloud computing scenarios. Whereas, a short loss of connection can lead to disastrous consequences for an autonomous vehicle system or an application impacting citizens' safety.

5.7 Networking, and storage enhancement

This is another big challenge for fog computing, especially after the emergence of software-defined environments such as SDN (Software-Defined Networking), SDHI (Software-Defined Hardware Infrastructures), VNF (Virtual Network Function), virtualization, SDC (Software-Defined Computing), SDI (Software-Defined Infrastructures), SDS (Software-Defined Storage), and others. Implementation of such techniques in fog networking requires a radical change in fog computing infrastructure design. It is not simple as it looks, but once it is done, all the other benefits especially latency minimization are achieved.

5.8 Energy consumption

By definition, the IoT objects collect and transmit data using wireless connections; fog computing also supports wireless D2D (Device to Device) connectivity, whereby the networks of devices can decrease significantly their energy consumption since a big amount of requests are executed in fog nodes. From another perspective, fog computing contributes to decrease cloud computing energy consumption because most of the IoT requests are onloaded to the border of the network.

5.9 Scalability

This feature is widely required in fog computing infrastructure. The fog data centers need to support the load balancing, agility, and elasticity of runtime, these variables contribute to efficiently control the variation in fog computing workload. This challenge is strongly linked to geo-distribution, since it has been often required for the fog data center to be efficiently geo-distributed, in that way each fog datacenter serves IoT devices existing in its coverage area. The need for scalability is triggered by the instant and high demand for the workload that can be created by IoT devices.

5.10 Complexity

It is obvious that IoT devices are limited in resources point of view, so onloading tasks to the fog layers reduce the computational complexity of IoT devices [66]. From another perspective, the fog/edge computing approach reduces network architecture complexity, as well as decreases the number of points of failures in IoT systems. Integrating ML capability in the fog layer minimizes the complexity of the decision-making process.

5.11 Heterogeneity

IoT architecture is becoming more heterogeneous day after day. A relevant definition of fog computing given by Yi et al. [67] mentioned that "*Fog Computing is a geographically distributed computing architecture with a resource pool which consists of one or more ubiquitously connected heterogeneous devices (including edge devices) ...*," this definition confirms that fog computing is supposed to manage data and devices coming from varied manufacturers. These devices have different physical characteristics and require a variety of deployment methods.

6. Blockchain technology for digital farming

We mean by Blockchain a digital and distributed ledger that protects the history of any digital asset from any alteration or unauthorized modification, this protection results from the use of hashing, cryptographic techniques, public-private key functions, distributed databases, and processing, as well as consensus algorithms. Blockchain is historically conceived in the creation of Bitcoin [68] by "*Satoshi Nakamoto*" in 2008 as a novel cryptocurrency purely digitalized. Although this technology is still in its early stage, it recently creates a technological revolution, thanks to its brought advantages, such as guaranteeing transparency, auditability, anonymity, decentralization [69], and independency, as well as reducing the risk of frauds in transactions between machines in a P2P (Peer to Peer) network. In the food and farming context, the combination between Blockchain and IoT is a promising contribution that immerged to improve the traditional methods of farming management. Blockchain technology can provide farmers with new solutions to smartly manage and monitor soil, engines, warehouses, livestock, logistics, and supply chain. It was expected in Ref. [70] that the utilization of this technology in the supply chain market will reach $ 429.7 million by 2023. The need to build trust between the food producers and the consumers in the agri-food sector is a big concern that can call Blockchain technology to provide transparency, efficiency, and sustainability in the agri-food chain. Moreover, the more quantity and diversity of food is produced, the more compliances and audits are required, the information resulting from the audits is still managed with traditional paper or stored in a centralized database, this approach of management is susceptible to suffer from many issues such as error, lack of integrity, and data consistency, as well as fraud and corruption in the case of paper-based information [71].

The following sub-sections discuss the possible solutions on how Blockchain technology can be used in digital farming and smart agriculture. Each section discusses some of the most relevant platforms adopted in Blockchain use cases upon which IoT-based intelligent farming applications are based. After consulting this sub-section, the reader will discover an obvious complementarity between the use cases, it is up to the implementer of the Blockchain-based application to decide either to combine many use cases in one system or to focus on one use case in its contribution. **Figure 4** illustrates the possible seven use cases of Blockchain in IF.

6.1 IoT security optimization

It is difficult for the traditional vision of networks to provide the requirements of IoT-based IF systems notably latency, bandwidth, security, and reliability. A Blockchain-based security architecture proposed to monitor the integrity of IoT collected data by checking and preventing unhallowed alteration that can be caused by DDoS (Distributed Denial of Service) attacks on delivered data [72]. The Blockchain-based solutions for improvement of IoT security in green agriculture cover many areas [73] such as public key infrastructure support [74], machine learning-based systems [75], access control improvement [76, 77], reputation and trust use case [78, 79], amelioration of authentication and identification of IoT objects thanks to the bubble of trust system [80]. The bubble of trust is analogically a private VLAN (Virtual Local Area Network) of sensors, communication between sensors in the same bubble is fully private and secured because it must be validated by the Blockchain network, furthermore, no communication out of this bubble is authorized. **Figure 5** shows a proposed scenario on how can Blockchain be applied to secure transactions in an IoT system. When the positioning system collects the location of the smart tractor, a transaction is occurred and is inserted in a new block, the generated block is sent to the other miners for checking the solution used in the mining process. Once the

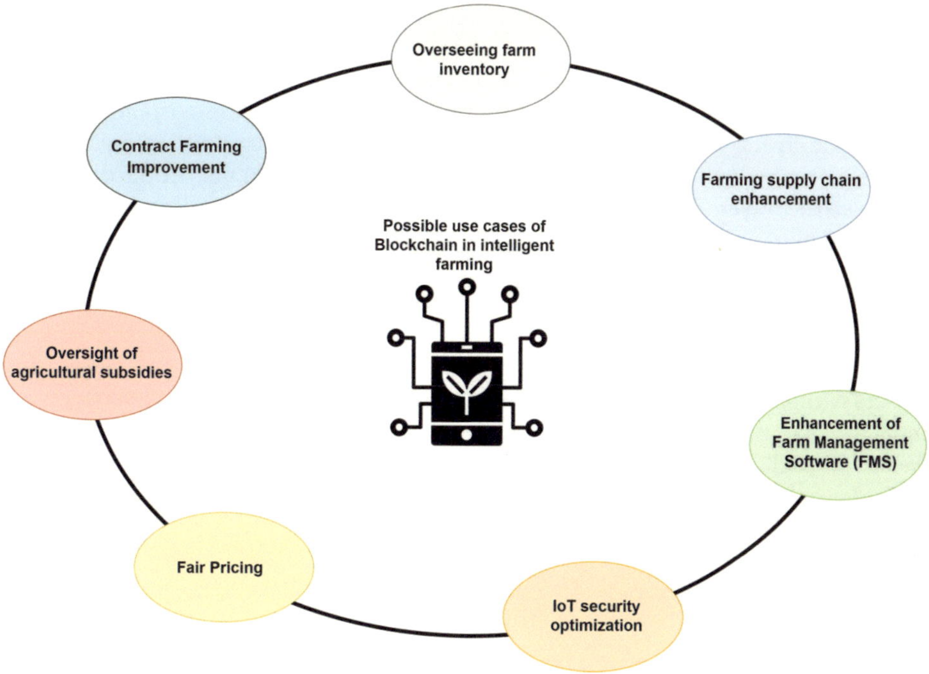

Figure 4.
The possible use cases of Blockchain in intelligent farming.

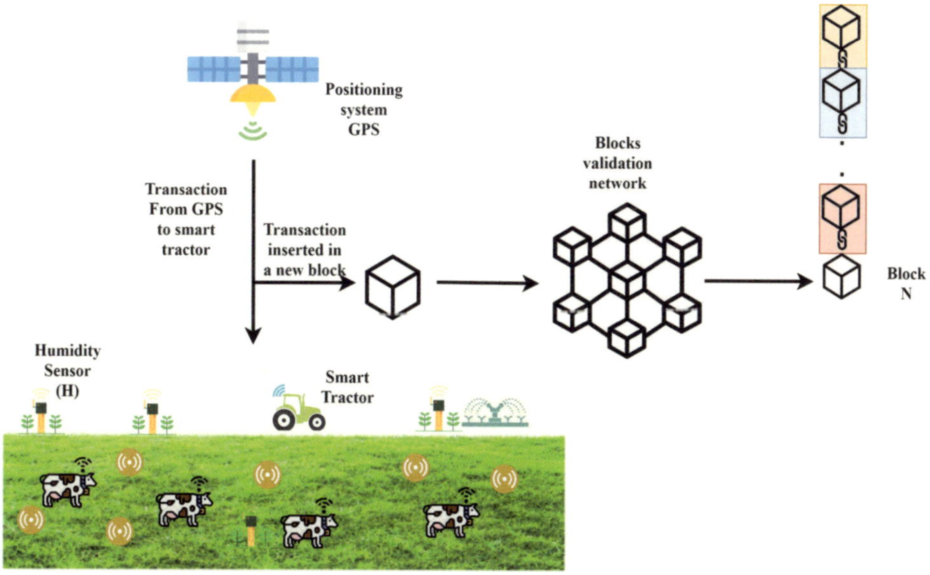

Figure 5.
A proposed scenario of a Blockchain-based IoT security optimization application.

mining solution is validated, the block is addressed to the Blockchain nodes for validation, and stored in the Blockchain once it is verified. This process is fully decentralized and uses cryptography techniques and hashing.

6.2 Fair pricing

Farmers are the weak link in the agri-food production chain, the price they got for their products does not reflect their real provided efforts due to the existence of multiple middle layers of buyers. This issue happens because they lack marketing opportunities, thus their products are not properly marketed, so they do not get the deserved price from the buyers. Thanks to Blockchain technology, farmers can reach more buyers and marketplaces than expected and can fairly discuss the right price of their goods. A decentralized farming approach named KHET is proposed by Paul et al. [81] to slightly reduce this issue, KHET platform enables farmers, companies, and buyers to communicate with each other, and make commitments based on the smart contract without any intermediary. With such a platform, farmers can finance their farming projects without requesting a loan from the bank. **Figure 6** illustrates a proposed model of how can farmers make deals fairly with retailers using Blockchain technology. The farmer and the retailers must be registered in the public Blockchain system, each one is identified with a unique identifier, which is its digital wallet address. The deals are made on a dedicated agricultural platform which is channeled with the Blockchain system using a dedicated API, the role of the API is to retrieve and verify farmers' and retailers' addresses. The farmers are now able to check and discuss the prices of their

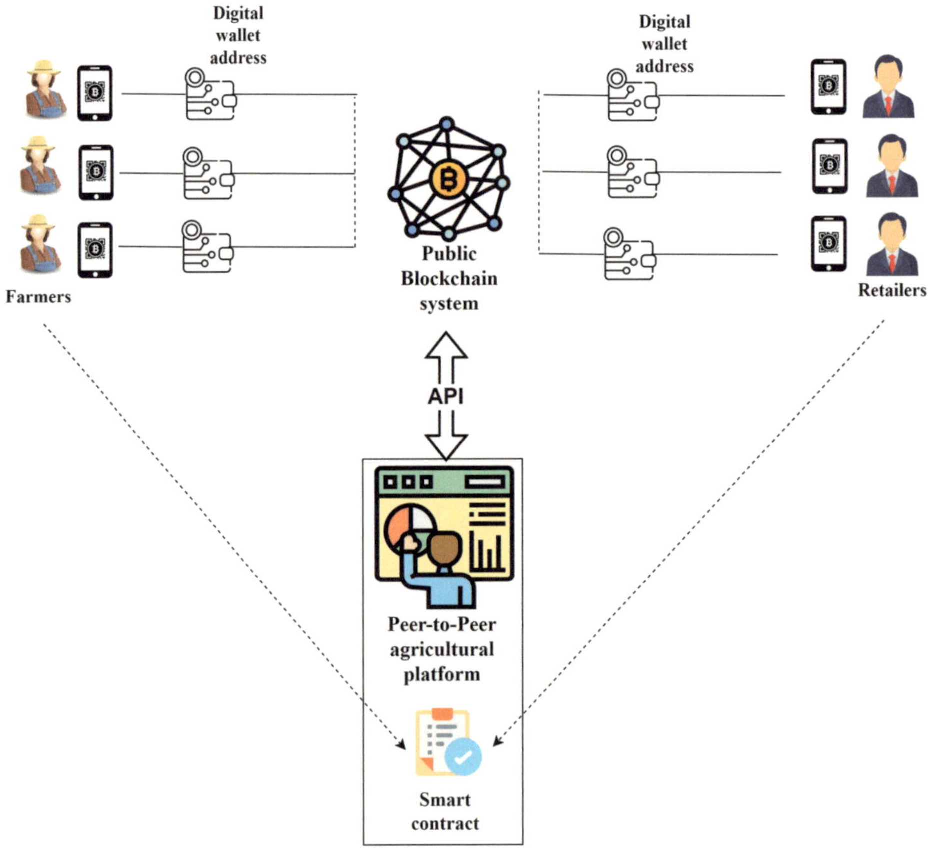

Figure 6.
A proposed model of agricultural fair pricing application based on Blockchain technology.

products freely and fairly with all interested stakeholders and without a middle-man. If the farmer and the retailer accept the conditions, the smart contract is established and the amount of money can also be transferred from the retailer's digital wallet to the farmer's digital wallet using the digital money platform.

6.3 Oversight of agricultural subsidies

To help farmers in their multiple investments and increase productivity, a new governmental subsidies distribution system should be adopted. The classical methods of distributing aids to farmers lack transparency due to information centralization and lack of coordination between agricultural stakeholders. With Blockchain, a decentralized ledger can be built to ensure agricultural information sharing in a secured manner. The digital ledger can be made publically available, thus farmers

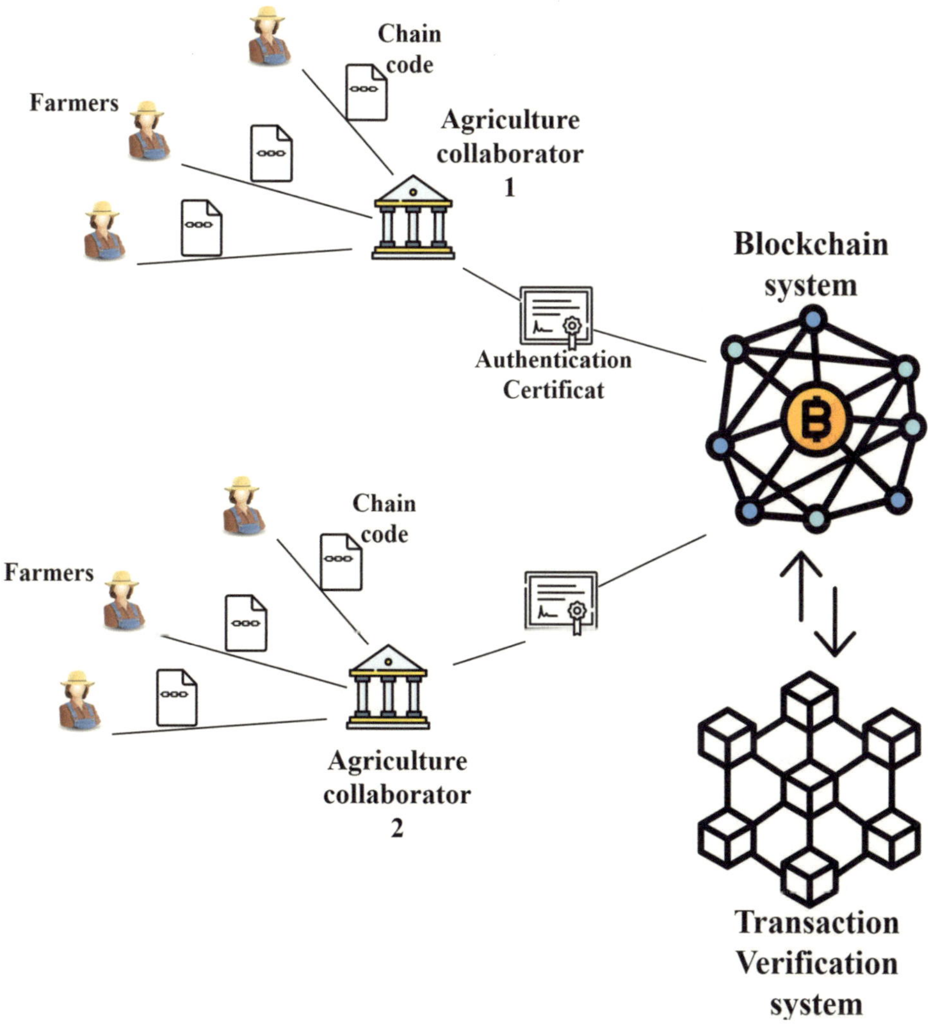

Figure 7.
A proposed scenario of single farmer identity management using Blockchain in a multi-collaborators environment.

can see if subsidies go it should be, as well as how much each farmer receives as aid. In this context, Abraham and Santosh Kumar [82] proposed a Blockchain-based system to ensure transparency and reliability of the information in the subsidies system. The scenario proposed in **Figure 7** provides a solution to deal with the problem of farmers' identity management in a multi-collaborators environment, each farmer is identified by a chain code which is a smart contract installed on the peers of the private system of the AD (Agricultural Department), each AD uses a certificate to authenticate the transaction in the public Blockchain system and keep a private validated ledger. When the farmer sends a transaction, it is accepted or refused depending on the rules and the policy described in the chain code. Agricultural departments are interfaced with the Blockchain system to share the information securely with each other using the unique identity of the farmer. When a transaction occurs between one or more AD, it must be validated by the transaction verification system, which is composed of the other agricultural collaborators. According to this scenario, farmers' information is transparent and reliable for all the agricultural collaborators, Thus, subsidies go to the one who deserves them.

6.4 Contract farming improvement

Smart contract occurs when it is self-managed without middle parties which increases automation and decentralization of the tamper-proof of data, Ethereum Blockchain [83] and Hyperledger Fabric represent an example of platforms that support this kind of technology. They allow developers to implement their Blockchain layer and applications, such as smart contracts, in a decentralized way. The Blockchain-based IF use case enables the final consumers and the partners to have full knowledge about the agricultural product that they want to buy or to retail. The integration of the smart contract with IoT by Umamaheswari et al. [84] helps to build trust between farmers and consumers by providing information about the origin and the environment in which the product is grown and stored, as well as the ability to track the transaction path. Moreover, the implementation of smart contract in the agricultural process improves the CIA (Confidentiality, Integrity, Availability) of data storing method and enable the public to get a trustable license based on the comparison between the products' stored information in the data private chain and those publically available [85]. Data sharing in the IF environment is one of the major challenges of the distributed and scalable IoT systems, this issue is managed by Ur Rahman et al. [86] through a data-sharing smart contract system with access control capability. The smart contract application is present in models proposed in **Figures 6-10**.

6.5 Overseeing farm inventory

Farmers work hard and wait for the post-harvest stage, it is difficult for a farmer to imagine any damage in quantity or quality of his produce. Massive quantities of agricultural products are wasted before it reaches the retailer. This big wastage can be avoided by monitoring some environmental parameters in the storage area. Humidity, temperature, and CO_2 concentration are some variables that can be tracked using IoT and sensors. Public ledgers using Blockchain allow to share information about the product storage operation between all the chain stakeholders, so big visibility about the product's history is provided to all interested collaborators. Moreover, combining IoT and sensors to gather information about the inventory, and public ledgers to implement strategies to monitor this information can be a perfect way

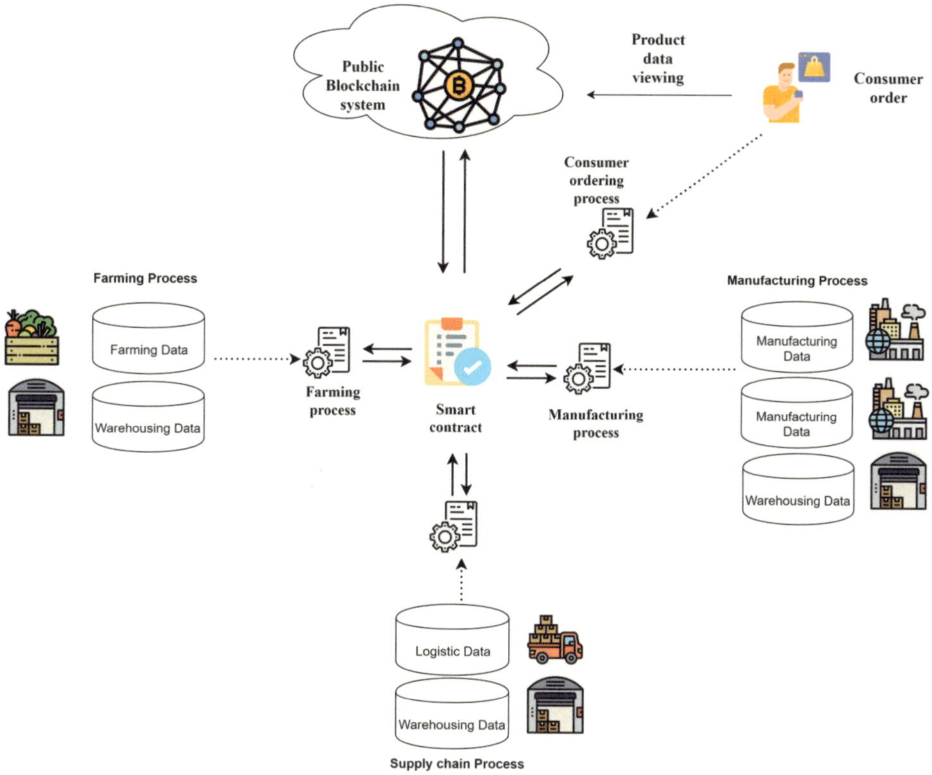

Figure 8.
A proposed scenario of overseeing farm inventory using Blockchain.

to manage inventories and logistics flows. Vendor-managed inventories (VMI) is a popular Blockchain-based collaborative inventory management policy, VMI might be founded on the smart contract between manufacturers, vendors, and buyers [87], consequently, each one of those collaborators can build its supply chain strategy and inventory policy management [88]. The proposed architecture in **Figure 8** illustrates a Blockchain-based system for product inventory management. Farming, manufacturing, and supply chain processes are authenticated using smart contracts and share the products' data in the Blockchain system publically available for consumers. All the transactions occurring between the consumer and the other stakeholders are managed and protected by the smart contract, the verified transaction are stored securely in the Blockchain. The consumer can check the information related to the products before ordering them, or track their safety on the farm, in the factory, or during the delivery process.

6.6 Farming supply chain enhancement

Demonstrating the quality of a product in a producer-consumer relationship is the critical weakness of community-supported agriculture [89]. Without transparency and mechanisms of tracking and monitoring in the production process, consumers are unsure about the safety of the goods they buy and receive. The traceability frameworks based on Blockchain technology in the supply chain is an important key feature not only to ensure the security of the on-chain or off-chain encrypted

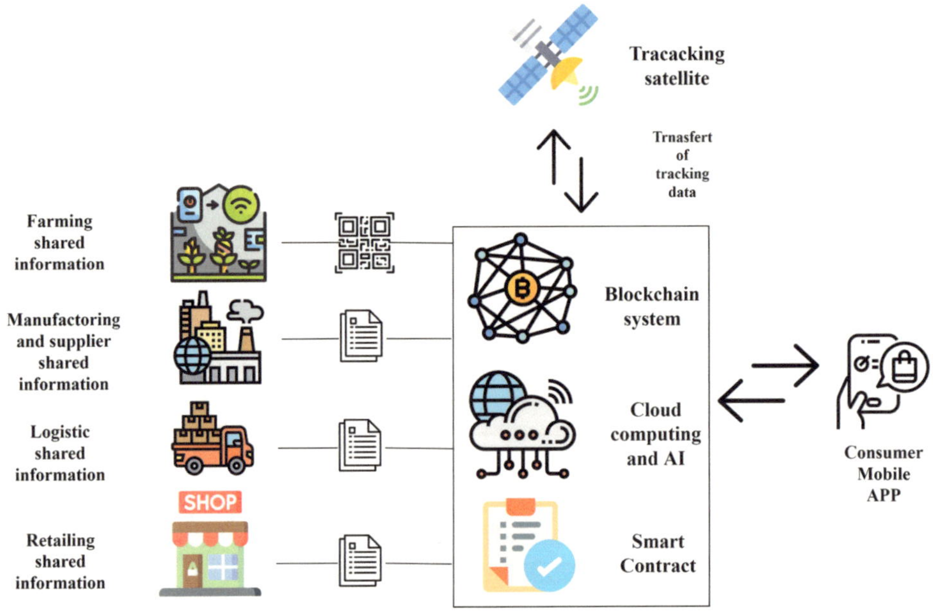

Figure 9.
A proposed model of supply chain enhancement using Blockchain technology.

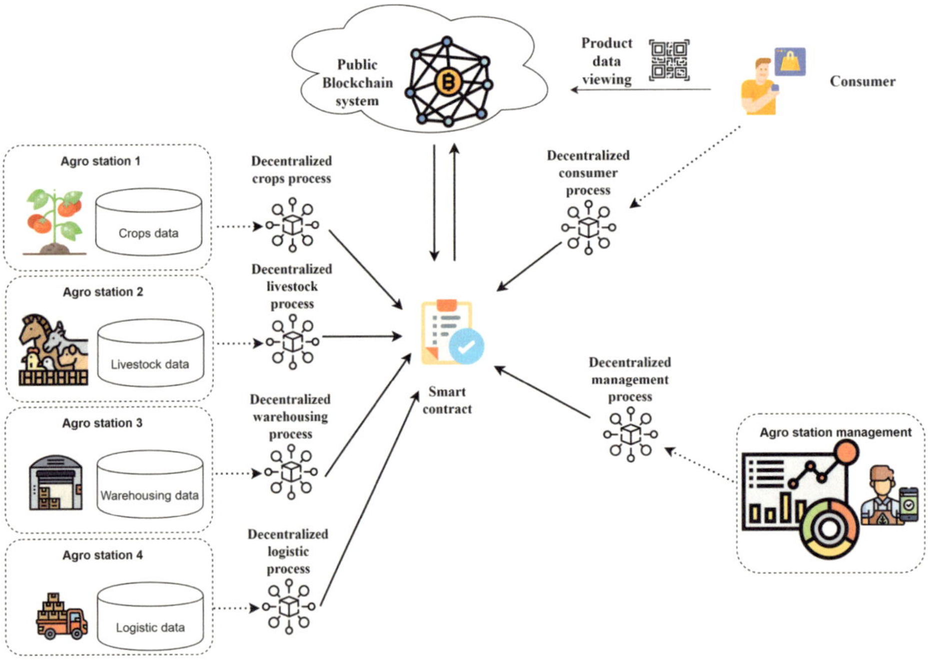

Figure 10.
A proposed Blockchain-based FMS scenario.

and stored data, but also to overcome the big latency that can be generated when querying databases [90] either by the public community or by the relevant partners. Combining IoT, RFID, and QR (Quick Response) code with Blockchain helps to build powerful supply chain systems to track agricultural food from farmer to retailer and make product information accessible to all users [91]. **Figure 9** shows a proposed model for a supply chain enhancement use case. The food information is shared in all the supply chain phases. IoT and sensors collect data related to the environment where the crop is grown, the manufacturing conditions, the shipment and logistic flow, and the retailing environment. The consumer through his mobile application generates a transaction (new command of a product) and checks the product's shared details. On the other hand, the supplier can make his offer, the smart contract is for protecting the valid transaction between consumer and supplier, as well as storing the new transactions in the Blockchain system.

6.7 Enhancement of farm management software

Modern farming requires the modernization of all its processes including FMS (Farming Management Software), traditional FMS are based on a classical client-server based-approach, this method does not satisfy the growing demand on inputs-outputs as well as enough security level for data protection. With Blockchain technology, more sophisticated and secured systems for supply chain management, smart greenhouse, and livestock are provided, so that farmers and analysts who care about data integrity and uncertainty will not worry anymore about intentional or accidental alterations that can be caused by one of the information flow manipulators. It is expected that the FMS market growth will reach $4.22 Billion by 2025 [92], thanks to the widespread of Blockchain solutions and the wide usage of IoT, sensors, as well as artificial intelligence in the farm management workflow. The model proposed in **Figure 10** explains an FMS use case. A secured and decentralized management of the farm's processes is achieved, the principal role of the smart contract is to authenticate all the decentralized processes and ensure the integrity of the transactions that can be occurred between them. The data gathered in each decentralized process are shared with the public consumers through the public Blockchain system, the consumer can check the origin, the expiry date, and other information related to the warehousing with a simple scan of the QR code of the product. If the consumer is satisfied, he/she can supply orders to the farmers, and the smart contract is established. The farm distributed processes and the consumers' orders are managed using the FMS decentralized consol.

7. Research and development in digital farming

An overview of the published literature on the actual status of ICT usage in digital farming, particularly IoT-fog/edge/cloud computing, and Blockchain technologies reveals that most growers are interested in understanding the optimum conditions in open-field and closed-field crop production that results in reducing inputs, and at the same time maximized crop yield and quality. Our previous studies and survey show that some of the trending research topics in this context include (1) development of digital twin models that receives live data from various wireless sensors for improving efficiency of crop production systems [93], (2) adaptation of multi-robot platforms for wireless and IoT data collection [94], (3) health assessment, stress identification, and early disease detection using UAV remote sensing [95], (4) development of soil-test kits that can be mounted on mobile-robots for spontaneous determination of macronutrients in soil [96], (5) yield prediction and yield estimation using model-based and AI algorithms [97–99], (6) evaluation of crop growth environment prior to the actual cultivation for preventing yield loss (i.e., predictive models that can be leveraged as a part of digital twin) [100], (7) development of virtual orchard models using photogrammetry [101], (8) smart irrigation with solar powered IoT controlled actuators [102], (9) reducing time losses of machinery and increasing their field efficiency by using fleet management software [103], and (10) robotic weeding and harvesting [104, 105]. The success of such systems in our point of view is intimately linked to some important factors like the accuracy and complexity of ML/DL algorithms used to make IF decisions, as well as the availability of enough datasets to train and validate the ML/DL algorithms. From a Blockchain point of view, the horizontal and vertical scalability of IoT systems introduces more complexity in data sharing models within IF systems. The success of Bitcoin, as a result of Blockchain, is proven but the mutual collaboration between Blockchain contributors requires more maturity. Moreover, more efforts and works have to be provided to sensitize the public, the community of regulators, and the contributors about the need to invest in Blockchain development, without forgetting to address the scalability challenge (technologically speaking, it has a direct impact on the number of transactions). Furthermore, farmers in IF ecosystems need to make payments and receive subsidies from the government using cryptocurrency, transactions in this situation are susceptible to be targeted with selfish mining [106]. Blockchain is an open system, any miner can join the chain, and selfish miners can outperform honest miners and then can threaten the security of the transaction. It is a fact that Blockchain frameworks and updates for coding are publicly available, but they often lack the needed level of validation and verification against bugs, security breaches, and errors [107], so new researches and efforts are required in this direction.

Another important needed research is how to achieve interoperability between the Blockchain projects namely cross-chain, or between Blockchain and the exiting data models. The required interoperability in Blockchain enables users to take the full benefits of distributed Blockchain in terms of sharing information smoothly. As the main purpose of Blockchain is to fight against the centralization aspect, a big concern should be given to show how to build a strategy to share agricultural data (known crops diseases and solutions, best practices to increase yield) between farmers' decentralized ecosystems. The environmental impact of these technologies is always ignored or never addressed. Since sensors and electromagnetic fields generated by gateways are directly interacting with animals, soil, and vegetation, a serious study should be made to evaluate the degree of impact that the waste material of such technologies can have on the environment.

7.1 Machine learning for IoT-based digital farming

The efficiency and effectiveness of agriculture are driven by machine learning and deep learning techniques, these two mechanisms enable machines to learn and analyze data without even being programmed. ML/DL has emerged simultaneously with the Big data discipline to detect relationships, analyze patterns, and make predictions in farming activities. An example of applying a supervised machine learning algorithm with multiple distance detection sensors for autonomous navigation of a field agent robot is proposed by the SunBot project and shown in **Figure 11**. This robot is used for health assessment inside berry orchards and to collect data for supporting digital agriculture. Since traditional approaches and methods for farming management do not allow to increase productivity, farms nowadays need to be partially or

Figure 11.
Application of machine learning as a knowledge-based control approach for assisted navigation of a four-wheel steering field robot agent. Source: SunBot.de.

fully automated using IoT systems to collect data, and ML/DL to make data inspections and drive the decision-making tasks. ML/DL technology helps farmers and scientists to select the appropriate species that respond to specific requirements in terms of diseases resistance, adaptation for specific aquatic or soil conditions, this classification task was quite tedious for farmers or scientists, but with ML/DL, a huge quantity of unorganized data is gathered and analyzed automatically to finally choose which genome is suitable for breeding. In some cases, such as plant health monitoring, it is needed to compare plants according to their colors, leaf morphology, and shapes, in that case, ML/DL can be the solution to perform the fast and accurate classification. In this context, Thaiyalnayaki et al. [108] used SVM to classify soybean diseases, and [109] performed plant leaf diseases classification based on visible symptoms.

Soil management is another farming process that has benefited from ML/DL and IoT technologies, the buried sensors collect real-time data about the underground ecosystems such as temperature and moisture, and transfer them to ML/DL algorithms to estimate the quantity of water needed for irrigation, or evaluate the quantity of nutrients required for optimal growth of crops. Superficial sensors play a major role in measuring temperature, humidity, pressure, evaporation, and evapotranspiration, these climatological and hydrological parameters among others can be used by ML/DL algorithms to estimate exactly how much water is needed to irrigate a given surface area without any wastage. To avoid wastage related to weather forecast uncertainty, Chen et al. [110] used a short-term weather forecasts method to propose an optimal irrigation strategy. Another important role of ML/DL in intelligent farming is the accurate yield prediction in quantity and quality, this prediction can be useful in crop monitoring tasks and market price forecasting. From this vision, many popular ML/DL algorithms are compared in Ref. [111] in terms of three crops yield prediction, they reported good prediction skills of the SVM ML algorithm compared to the other tested ML/DL methods. Traditional methods to control crops diseases widely spread pesticides in all the field, this treatment method leads to wastage and does not ensure the required level of efficiency, as well as harming of environment. Modern farms use computer vision techniques to accurately detect where to apply pesticides, when to apply, how much is needed, and use drones to apply pesticides with high precision. Consequently, more financial benefits are won by the farmer with no environmental side effects. Weeds density detection and treatment are examples of computer vision use case that was applied by [112] to control the area of treatment.

Like crops management monitoring, there is livestock management monitoring, the use of IoT and ML/DL in this farming activity enables farmers to predict the productivity of meets and eggs based on actual or past data. For example, a drone can make a scan of the field and count the number and the position of the cattle. A computer vision system with smart cameras can monitor the mental condition of cows to detect their preferred time of milking or the quantity of feeds they want, as well as the amount of nutrients in their milk using sensors. The visible symptoms detected through computer vision techniques are used to measure animal welfare by monitoring the health conditions of animals, and predicting if a member of the cattle is sick or wants to eat or to drink.

7.2 Wireless communication for seamless connectivity in digital farming

Connectivity, as we said earlier, is an important component in IoT smart systems, this component is a challenging issue in rural environments where cellular network coverage may be absent, or only 2G networks are available, in this kind of cellular network, a limited number of devices can be supported that leads to a lack or reduced performance in data transfer. Nowadays, 3G/4G cellular networks are enough to build usual and smart farming applications. However, to unlock the potentials of IoT systems, two promising connectivity solutions, according to McKinsey Global Institute [113], are expected to be developed, these technologies are being referred to as "advanced" and "frontier." An example includes IoT-based collision avoidance sensors for autonomous electrical mowers that are capable of transmitting their distance measurement via WiFi and LoRa. While the main communication between different

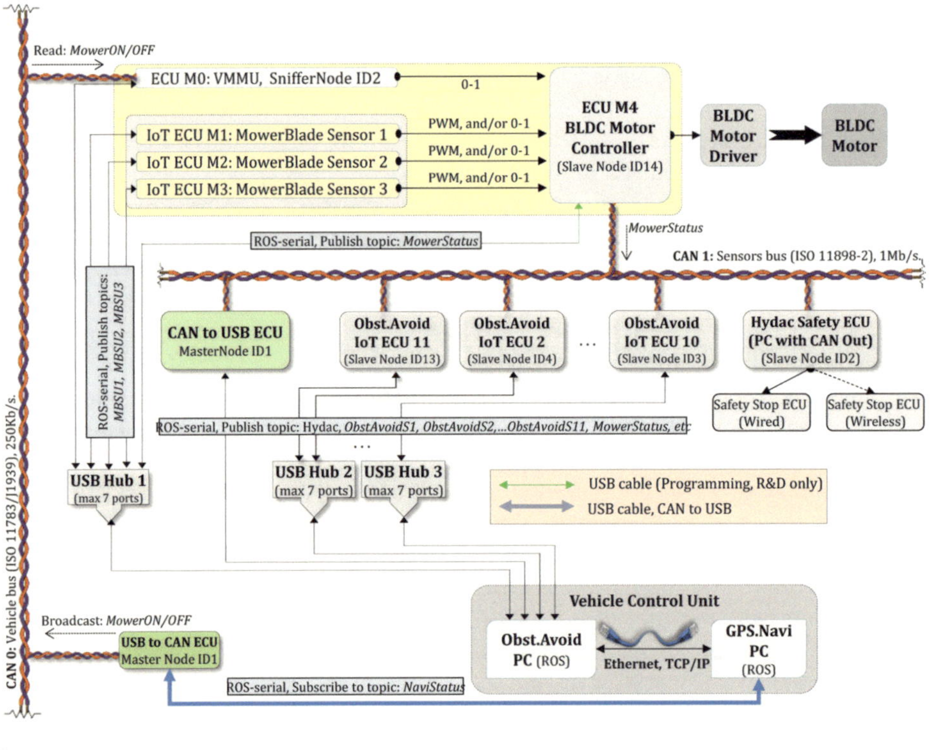

Figure 12.
Perception system with IoT-based LPWAN sensors for collision avoidance of a robotic mower. Source: SunBot.de.

electrical control units (ECU) for such system still relies on CANBUS and the detected distances can be logged on an onboard SD card (**Figure 12**), but the use of IoT-based ECUs that are independent of GPS and WiFi, provide the operator with LoRa messages for real-time monitoring of the mower status. This approach also makes possible simple switch control of the device in remote areas where WiFi and mobile coverage is not available. The architecture of this system is shown in **Figure 12**.

The advanced connectivity represents the next generation of already existing infrastructures, we mention here the upgrade that is occurring by providers of 4G technology toward 5G, this upgrade offers more improvement in speed, bandwidth, and latency, and the number of supported devices will be increased as well. For now, the evolution of wired connectivity, such as optical fibers, can offer the best performances in terms of latency, bandwidth, and speed especially in the core of the network, or in environments where mobility is not a crucial factor. Not Far from wireless networks, the Wi-Fi Alliance has certified the new standard 802.11ax known as Wi-Fi 6/6Extended, this new connectivity solution offers for devices a wide range of frequency and improved gain of speed that was estimated to achieve 40%, the theoretical speed of the network was estimated to reach 10 Gb/s, the Wi-Fi 6E offers 11 Gb/s as a theoretical speed with larger spectrum channels. These advantages enable IF devices to be connected seamlessly and smoothly, and the number of supported devices will be improved as well. The revolution in connectivity solutions has also been made by short-range technologies (Bluetooth, Wi-Fi, RFID) and low power wide area networks (LPWAN, LoRa, LoRaWan, NB-IoT), these technologies are usually used for tagging, tracking, or identification. These technologies have become more sophisticated and adapted for seamless connectivity in intelligent farming. The frontier connectivity is mostly designed for high mobility systems that need high speed, reliability, security, and minimal latency. Low earth orbit (LEO) and 5G networks are two options that will be developed to satisfy all IoT requirements. LEO constellations provide seamless connectivity services for IoT-based IF systems installed in distributed rural areas, or in zones where the terrestrial network is not available, so satellite coverage is needed. The other option of frontier connectivity is the 5G cellular networks, which promises to combine all the advantages of wired fiber in the air to be more adapted to IoT systems and wireless sensor networks.

7.3 Connectivity challenges of wireless sensing under field conditions

In remote areas, it is more adapted to use wireless devices as they allow to cover wider areas, but the energy consumed by these devices and their limited source of energy creates a big challenge that needs to be addressed. **Figure 13** shows multiple solar-powered LoRa sensors that have been deployed in different berry orchards in the state of Brandenburg in Germany for IoT monitoring of agricultural parameters (i.e., air and soil temperature, relative humidity, soil moisture, leaf wetness, light condition, and dew-point temperature). The wider area the IF system covers, the more power is consumed, some solutions are proposed to solve this issue, such as photovoltaic panels and the choice of low power consumption sensors. For instance, if BLE or low power consumption devices are used, the coverage area will be reduced because energy consumption will also be reduced, but if a wider communication range is needed, Wi-Fi connectivity can be adopted but energy consumption will be high. Technologies like LPWAN, LoRa, and LoRaWan adopt more efficient energetic strategies and a high communication range. Another connectivity limitation is the wireless signal quality. In remote areas where geographical issues are encountered, the

Figure 13.
Implementation of multiple solar-powered LoRa sensors in different berry orchards for IoT monitoring of field parameters. Source: SunBot.de.

wireless signal may have an attenuation problem because of multiple environmental obstacles or electromagnetic noises that can be introduced. The propagation of wireless signals can also be an issue that can be mitigated by installing signal repeaters or designing more efficient topologies such as mesh. The IoT and WSN systems management is another solution to reduce the connectivity limitations of intelligent farming systems, some of the management best practices are: (1) Designing an optimal size of the sensor network, here the number of sensors and the number of intermediary nodes to reach the gateway are to be considered because this factor impacts the communication range and the latency of data transmission. (2) The calibration of all WSN nodes whether sensors or gateways, this maintenance action improves the lifetime of the battery, especially in devices that operate in a wide range [114]. (3) Using optimized transmission protocols, many protocols are identified in the literature as efficient solutions to optimize transmission tasks, either to save the energy of the battery, to optimize the routing strategy, or to increase the coverage area.

7.4 Challenges with IoT monitoring in remote areas

Other issues that are encountered when designing an IoT-based intelligent farming system are related to interoperability [115], technological development, data heterogeneity management, scalability and flexibility of the system, fault tolerance, complexity of the system and the harsh environment, energetic issue, and the need for professionals to implement and manage the system. The interoperability issue takes four different formats, it can be technical, organizational, semantic, or synthetical, all of these four components are interdependent, but the most common issue is the technical one, this is occurred due to the hardware and software differences between manufacturers, these differences imply heterogeneity in protocols and connectivity standards, so when implementing the IF system, the farmer finds himself in front of many incompatible technical choices that he should manage particularly if there is an already existing system that it has to be taken into account. The integration issue can go beyond hardware compatibility to software conflicts that can create a new challenge of integrating new IoT points with the existing management software or vice-versa. The velocity of technological development is another issue of IoT implementation in IF, the hardware and the software related to IoT systems are evolving rapidly, which leads to the continuous emergence of new efficient frameworks, the upgrade process can be expensive in terms of infrastructure or maintenance. The scalability and flexibility of the IF system measure the level of opening, centralization, ease of integration with other existing systems and platforms, and ability to scale the system in terms of the number of nodes and storage, this issue represents an example of organizational interoperability. We rarely find all the implemented components of the IoT system from the same manufacturer, this technological heterogeneity and the lack of a global standard that unifies the format of data managed by each technology is challenging for the farmer. Some efforts in this context have been made by the Agricultural Industry Electronics Foundation (AEF) to propose the ISOBUS database (actual version is ISO 11783-1:2017) as an attempt to fill the heterogeneity in data format for agricultural machinery, this issue represents an example of semantic interoperability. The fault-tolerance measures the robustness of the designed IF system. When implementing the IoT-based IF system, the farmer is invited to manage all the hardware faults and system errors that can be occurred, the fewer harmful events the system generates, the more reliability the system has. However, farmers need to have particular skills for better management of these damaging events. As we discussed before, the power strategy in IF systems represents a big issue that makes energetic barriers in front of IoT systems implementation and needs to be taken into account. Because the farming system is composed of multiple heterogeneous hardware and software components, the management and the integration tasks could be more or less difficult depending on the level of complexity generated by the adopted topology, the interoperability between the elements of the system, and the opening degree of the adopted technology. In fact, the complexity is not an issue for the farmer only, but the manufacturers also should consider it while designing their products. The reliability and efficiency of the IF system are greatly impacted by the environment where it is deployed, geographical and climatological characteristics such as high temperature, wind speed, heavy rain, and dusty environments can destroy the sensors or can make them totally out of service [116]. Thus, choosing the hardware that resists environmental damages is considered a big responsibility that should be

Figure 14.
Redundant LoRa sensors with modular accessories and multiple transmitters and gateways to overcome uncertainties and connectivity issues in actual field conditions. Source: SunBot.de.

considered when implementing the IoT-based IF system. **Figure 14** shows a modular IoT solution with multiple LoRa sensors and gateways that have been custom-built for the SunBot project to withstand harsh field conditions and overcome the issues with WiFi instability. Each sensor is benefitting from multiple transmitters to reduce the probability of signal loss, and multiple gateways to ensure data uploads to the private cloud.

8. Conclusion

The interactions between the human and virtual world are increasingly developing day after day, thanks to the widespread connectivity solutions and the ubiquity of connected objects that rapidly become smart. ML/DL also is one of the promising topics that gain recently the big attention of the research community since it capitalizes the efforts made in IoT data management fields and the evolution of Fog/cloud computing paradigms. In this survey, we discussed the IoT-based systems' requirements and shed light on the components of an intelligent farming IoT model as well as the open challenges resulting from the integration of IoT systems and fog computing technology. We talked later about Blockchain technology, its applications to improve the intelligence and the security of the farming field. From another hand, we discussed the needed researches to apply Blockchain more accurately in the farming domain. This paper is closed with a discussion about the main limitations that the implementation of IoT in intelligent farming is facing. In summary, the significant results of this survey can be summarized in the three following points—(1) this survey investigates the implementation of ICT in farming environments to solve many current serious issues related to management methods. IoT-based applications combined with machine learning are complete solutions to efficiently improve crop yields without wasting too much resources. The second result concerns Blockchain technology that can be integrated with IoT-based farming systems to provide efficient security solutions and build trust between farmers each other, or between farmers and consumers. Furthermore, we enable the reader to discover the seven significant applications of Blockchain in the intelligent farming field to improve security in IoT systems, fair pricing, agricultural subsidies oversight, the smart contract to securely manage the relationships between all the farming stakeholders, farm inventory overseeing, amelioration of supply chain and farm management software. This study also summarizes the open challenges resulting from the integration of IoT with fog/edge mining that creates many research problematics as well as makes the implementation of such solutions in the farming world very challenging tasks. (2) Many previous papers addressed the issue of implementing ICT in farming processes, but this work particularly elaborated the transition from cloud computing to fog/edge computing to serve IoT applications and added the integration of Blockchain in the farming field, its benefits, challenges, and applications. Finally, some recommended researches are needed to concretize the implementation of the proposed Blockchain models and propose another model for each farming activity. From another hand, the development of Blockchain technology requires serious investment efforts to provide a complete legal arsenal for better and safe implementation. (3) Although Blockchain technology is designed to build trust, its implementation in the intelligent farming workflow is still confronting many barriers related to the lack of trust [117] notably regulatory uncertainty (with 48%), lack of trust among users (45%), separate Blockchain systems not working together (41%), inability to scale (21%), intellectual property concerns (30%), and audit-compliance concerns (20%).

Author details

Mohammed El Idrissi[1*], Omar El Beqqali[1], Jamal Riffi[1], Redmond R. Shamshiri[2], Sanaz Shafian[3] and Ibrahim A. Hameed[4]

1 Faculty of Science D.M., Sidi Mohamed Ben Abdellah University, Atlas-Fez, Morocco

2 Leibniz Institute for Agricultural Engineering and Bioeconomy, Potsdam-Bornim, Germany

3 School of Plant and Environmental Sciences, Virginia Polytechnic Institute and State University, Blacksburg, USA

4 Faculty of Information Technology Department of ICT and Natural Sciences, and Electrical Engineering, NTNU, Ålesund, Norway

*Address all correspondence to: mohammed.elidrissi29@usmba.ac.ma

References

[1] Karlsen KM, Dreyer B, Olsen P, Elvevoll EO. Literature review: Does a common theoretical framework to implement food traceability exist? Food Control. 2013;**32**:409-417. DOI: 10.1016/j.foodcont.2012.12.011

[2] Faisal MN, Talib F. Implementing traceability in Indian food-supply chains: An interpretive structural modeling approach. Journal of Foodservice Business Research. 2016;**19**:171-196. DOI: 10.1080/15378020.2016.1159894

[3] Haleem A, Khan S, Khan MI. Traceability implementation in food supply chain: A grey-DEMATEL approach. Information Processing in Agriculture. 2019;**6**:335-348. DOI: 10.1016/j.inpa.2019.01.003

[4] Khan S, Haleem A, Khan M, Abidi M, Al-Ahmari A. Implementing traceability systems in specific Supply Chain Management (SCM) through critical success factors (CSFs). Sustainability. 2018;**10**:204. DOI: 10.3390/su10010204

[5] García L, Parra L, Jimenez JM, Lloret J, Lorenz P. IoT-based smart irrigation systems: An overview on the recent trends on sensors and IoT systems for irrigation in precision agriculture. Sensors. 2020;**20**:1042. DOI: 10.3390/s20041042

[6] Taneja M, Jalodia N, Byabazaire J, Davy A, Olariu C. SmartHerd management: A microservices-based fog computing–assisted IoT platform towards data-driven smart dairy farming. Software: Practice and Experience. 2019;**49**:1055-1078. DOI: 10.1002/spe.2704

[7] Smith D, Rahman A, Bishop-Hurley GJ, Hills J, Shahriar S, Henry D, et al. Behavior classification of cows fitted with motion collars: Decomposing multi-class classification into a set of binary problems. Computers and Electronics in Agriculture. 2016;**131**:40-50. DOI: 10.1016/j.compag.2016.10.006

[8] Ratnaparkhi S, Khan S, Arya C, Khapre S, Singh P, Diwakar M, et al. Smart agriculture sensors in IOT: A review. Materials Today: Proceedings. 2020. DOI: 10.1016/j. matpr.2020.11.138

[9] Tahsien SM, Karimipour H, Spachos P. Machine learning based solutions for security of Internet of Things (IoT): A survey. Journal of Network and Computer Applications. 2020;**161**:102630. DOI: 10.1016/j.jnca.2020.102630

[10] Hajjaji Y, Boulila W, Riadh Farah I, Romdhani I, Hussain A. Big data and IoT-based applications in smart environments: A systematic review. Computer Science Review. 2021;**39**:100318. DOI: 10.1016/j.cosrev.2020.100318

[11] Farooq MS, Riaz S, Abid A, Abid K, Naeem MA. A survey on the role of IoT in agriculture for the implementation of smart farming. IEEE Access. 2019;**7**:156237-156271. DOI: 10.1109/access.2019.2949703

[12] Mekala MS, Viswanathan P. A survey: Smart agriculture IoT with cloud computing. In: Proceedings of the International Conference on Microelectronic Devices, Circuits and Systems (ICMDCS); 10-12 August 2017; India. New York: IEEE; 2017. pp. 1-7

[13] Cisternas I, Velásquez I, Caro A, Rodríguez A. Systematic literature review

of implementations of precision agriculture. Computers and Electronics in Agriculture. 2020;**176**:105626. DOI: 10.1016/j.compag.2020.105626

[14] Lova Raju K, Vijayaraghavan V. IoT technologies in agricultural environment: A survey. Wireless Personal Communications. 2020;**113**:2415-2446. DOI: 10.1007/s11277-020-07334-x

[15] Shi X, An X, Zhao Q, Liu H, Xia L, Sun X, et al. State-of-the-art internet of things in protected agriculture. Sensors. 2019;**19**:1833. DOI: 10.3390/s19081833

[16] Muangprathub J, Boonnam N, Kajornkasirat S, Lekbangpong N, Wanichsombat A, Nillaor P. IoT and agriculture data analysis for smart farm. Computers and Electronics in Agriculture. 2019;**156**:467-474. DOI: 10.1016/j.compag.2018.12.011

[17] Bacco M, Barsocchi P, Ferro E, Gotta A, Ruggeri M. The digitisation of agriculture: A survey of research activities on smart farming. Array. 2019;**3-4**:100009. DOI: 10.1016/j.array.2019.100009

[18] Cees Links. The Internet of Things Will Change our World, ERCIM News101[Internet]. 2015. Available from: https://ercimnews.ercim.eu/images/stories/EN101/EN101-web.pdf [Accessed: June 24, 2020]

[19] Ray PP. A survey on internet of things architectures. Journal of King Saud University—Computer and Information Sciences. 2018;**30**:291-319. DOI: 10.1016/j.jksuci.2016.10.003

[20] Yi C, Cai J, Su Z. A multi-user mobile computation offloading and transmission scheduling mechanism for delay-sensitive applications. IEEE Transactions on Mobile Computing. 2020;**16**:29-43. DOI: 10.1109/tmc.2019.2891736

[21] Bhardwaj K, Shih MW, Agarwal P, Gavrilovska A, Kim T, Schwan K. Fast, scalable and secure onloading of edge functions using airbox Edge Computing. In: Proceedings of the IEEE/ACM Symposium on Edge Computing (SEC); 27-28 October 2016; USA. New York: IEEE; 2016. pp. 14-27

[22] Yang L, Cao J, Cheng H, Ji Y. Multi-user computation partitioning for latency sensitive mobile cloud applications. IEEE Transactions on Computers. 2015;**64**:2253-2266. DOI: 10.1109/tc.2014.2366735

[23] Yousefpour A, Ishigaki G, Gour R, Jue JP. On reducing IoT service delay via fog offloading. IEEE Internet of Things Journal. 2018;**5**:998-1010. DOI: 10.1109/jiot.2017.2788802

[24] Molina M, Munoz O, Pascual-Iserte A, Vidal J. Joint scheduling of communication and computation resources in multiuser wireless application offloading. In: Proceedings of the IEEE Annual International Symposium on Personal, Indoor, and Mobile Radio Communication (PIMRC); 2-5 September 2014; USA. New York: IEEE; 2015. pp. 1093-1098

[25] Ren J, Yu G, He Y, Li Y. Collaborative cloud and edge computing for latency minimization. IEEE Transactions on Vehicular Technology. 2019;**68**:5031-5044. DOI: 10.1109/TVT.2019.2904244

[26] Xiang X, Lin C, Chen X. Energy-efficient link selection and transmission scheduling in mobile cloud computing. IEEE Wireless Communications Letters. 2014;**3**:153-156. DOI: 10.1109/wcl.2013.122113.130825

[27] Ge Y, Zhang Y, Qiu Q, Lu YH. A game theoretic resource allocation for overall energy minimization in mobile cloud computing system. In: Proceedings of the ACM/IEEE International Symposium

on Low Power Electronics and Design (ISLPED'12); July 2012; USA. New York: Association for Computing Machinery; 2012. pp. 279-284

[28] Chen X, Jiao L, Li W, Fu X. Efficient multi-user computation offloading for mobile-edge cloud computing. IEEE/ACM Transactions on Networking. 2016;**24**:2795-2808. DOI: 10.1109/tnet.2015.2487344

[29] Mostafa N, Ridhawi IA, Aloqaily M. Fog resource selection using historical executions. In: Proceedings of the International Conference on Fog and Mobile Edge Computing (FMEC); 23-26 April 2018; Spain. New York: IEEE; 2018. pp. 272-276

[30] Jana GC, Banerjee S. Enhancement of QoS for fog computing model aspect of robust resource management. In: Proceedings of the International Conference on Intelligent Computing, Instrumentation and Control Technologies (ICICICT); 6-7 July 2017; India. New York: IEEE; 2018. pp. 1462-1466

[31] Souza VBC, Ramirez W, Masip-Bruin X, Marin-Tordera E, Ren G, Tashakor G. Handling service allocation in combined Fog-cloud scenarios. In: Proceedings of the IEEE International Conference on Communications (ICC); 22-27 May 2016; Malaysia. New York: IEEE; 2016. pp. 1-5

[32] Aazam M, Huh EN. Dynamic resource provisioning through fog micro datacenter. In: Proceedings of the IEEE International Conference on Pervasive Computing and Communication Workshops (PerCom); 23-27 March 2015; USA. New York: IEEE; 2015. pp. 105-110

[33] Das ML. Privacy and security challenges in internet of things. In: Natarajan R, Barua G, Patra MR, editors. Distributed Computing and Internet Technology. ICDCIT 2015. Cham: Springer; 2015. pp. 33-48. DOI: 10.1007/978-3-319-14977-6_3

[34] Hernández-Ramos JL, Pawlowski MP, Jara AJ, Skarmeta AF, Ladid L. Toward a lightweight authentication and authorization framework for smart objects. IEEE Journal on Selected Areas in Communications. 2015;**33**:690-702. DOI: 10.1109/jsac.2015.2393436

[35] Zhang P, Chen Z, Liu JK, Liang K, Liu H. An efficient access control scheme with outsourcing capability and attribute update for fog computing. Future Generation Computer Systems. 2018;**78**:753-762. DOI: 10.1016/j.future.2016.12.015

[36] Chaisiri S, Lee BS, Niyato D. Optimization of resource provisioning cost in cloud computing. IEEE Transactions on Services Computing. 2012;**5**:164-177. DOI: 10.1109/tsc.2011.7

[37] Prasad A, Lunden P, Moisio M, Uusitalo MA, Li Z. Efficient mobility and traffic management for delay tolerant cloud data in 5G networks. In: Proceedings of the IEEE Annual International Symposium on Personal, Indoor, and Mobile Radio Communications (PIMRC); 30 August-2 September 2015; China. New York: IEEE; 2015. pp. 1740-1745

[38] Ning Z, Huang J, Wang X. Vehicular Fog computing: Enabling real-time traffic management for smart cities. IEEE Wireless Communications. 2019;**26**:87-93. DOI: 10.1109/mwc.2019.1700441

[39] Jayaraman PP, Palmer D, Zaslavsky A, Georgakopoulos D. Do-it-Yourself Digital Agriculture applications with semantically enhanced IoT platform. In: Proceedings of the IEEE International Conference on Intelligent Sensors, Sensor Networks and Information Processing

(ISSNIP); 7-9 April 2015; Singapore. New York: IEEE; 2015. pp. 1-6

[40] Desai P, Sheth A, Anantharam P. Semantic gateway as a service architecture for IoT interoperability. In: Proceedings of the IEEE International Conference on Mobile Services (MS); 27 June-2 July 2015; USA. New York: IEEE; 2015. pp. 313-319

[41] Ullah F, Habib MA, Farhan M, Khalid S, Durrani MY, Jabbar S. Semantic interoperability for big-data in heterogeneous IoT infrastructure for healthcare. Sustainable Cities and Society. 2017;**34**:90-96. DOI: 10.1016/j.scs.2017.06.010

[42] Tseng CL, Lin FJ. Extending scalability of IoT/M2M platforms with Fog computing. In: Proceedings of the IEEE World Forum on Internet of Things (WF-IoT); 5-8 February 2018; Singapore. New York: IEEE; 2018. pp. 825-830

[43] Vilalta R, Lopez V, Giorgetti A, Peng S, Orsini V, Velasco L, et al. TelcoFog: A unified flexible fog and cloud computing architecture for 5G networks. IEEE Communications Magazine. 2017;**55**:36-43. DOI: 10.1109/mcom.2017.1600838

[44] Gupta H, Nath SB, Chakraborty S, Ghosh SK. SDFog: A software defined computing architecture for QoS aware service orchestration over edge devices. arXiv 2016. Preprint arXiv:1609.01190

[45] Ito Y, Koga H, Iida K. A bandwidth allocation scheme to meet flow requirements in mobile edge computing. In: Proceedings of the IEEE International Conference on Cloud Networking (CloudNet); 25-27 September 2017; Czech Republic. New York: IEEE; 2017. pp. 1-5

[46] Gia TN, Jiang M, Rahmani AM, Westerlund T, Liljeberg P, Tenhunen H. Fog computing in healthcare internet of things: A case study on ecg feature extraction. In: Proceedings of the IEEE International Conference on Computer and Information Technology; Ubiquitous Computing and Communications; Dependable, Autonomic and Secure Computing; Pervasive Intelligence and Computing (CIT/IUCC/ DASC/PICOM); 26-28 October 2015; UK. New York: IEEE; 2015. pp. 356-363

[47] Roozbeh A, Soares J, Maguire GQ, Wuhib F, Padala C, Mahloo M, et al. Software-defined "hardware" infrastructures: A survey on enabling technologies and open research directions. IEEE Communications Surveys & Tutorials. 2018;**20**:2454-2485. DOI: 10.1109/comst.2018.2834731

[48] Ameen RY, Hamo AY. Survey of server virtualization. arXiv 2013. Preprint arXiv:1304.3557

[49] Blenk A, Basta A, Reisslein M, Kellerer W. Survey on network virtualization hypervisors for software defined networking. IEEE Communications Surveys & Tutorials. 2016;**18**:655-685. DOI: 10.1109/comst.2015.2489183

[50] Bruno Chatras, ETSI NFV Chair. Network Functions Virtualisation [Internet]. 2021. Available from: https://www.etsi.org/technologies/nfv [Accessed: June 25, 2021]

[51] Yasrebi P, Bemby S, Bannazadeh H, Leon-Garcia A. VNF service chaining on SAVI SDI. In: Atanasovski V, Leon-Garcia A, editors. Future Access Enablers for Ubiquitous and Intelligent Infrastructures. FABULOUS 2015. Cham: Springer; 2015. pp. 11-17. DOI: 10.1007/978-3-319-27072-2_2

[52] Buyya R, Calheiros RN, Son J, Dastjerdi AV, Yoon Y. Software-defined cloud computing: Architectural

elements and open challenges. In: Proceedings of the International Conference on Advances in Computing, Communications and Informatics (ICACCI); 24-27 September 2014; India. New York: IEEE; 2014. pp. 1-12

[53] Jararweh Y, Al-Ayyoub M, Darabseh A, Benkhelifa E, Vouk M, Rindos A. Software defined cloud: Survey, system and evaluation. Future Generation Computer Systems. 2016;**58**:56-74. DOI: 10.1016/j.future.2015.10.015

[54] Ashton K. That 'internet of things' thing. RFID Journal. 2009;**22**:97-114

[55] Shariatmadari H, Ratasuk R, Iraji S, Laya A, Taleb T, Jäntti R, et al. Machine-type communications: Current status and future perspectives toward 5G systems. IEEE Communications Magazine. 2015;**53**:10-17. DOI: 10.1109/mcom.2015.7263367

[56] Al-qudah Z. Optical wireless communications: Current status and future prospects. Innovative Systems Design and Engineering. 2017;**8**:25-30

[57] Qadir QM, Rashid TA, Al-Salihi NK, Ismael B, Kist AA, Zhang Z. Low power wide area networks: A survey of enabling technologies. Applications and Interoperability Needs. IEEE Access. 2018;**6**:77454-77473. DOI: 10.1109/access.2018.2883151

[58] Sigfox. [Internet]. 2019. Available from: http://www.sigfox.com [Accessed: September 09, 2019]

[59] 3GPP TSG GERAN 65 [Internet]. Combined Narrow-Band and Spread Spectrum Physical Layer Coverage and Capacity Simulations. 2015. Available from: https://www.3gpp.org/ftp/tsg_geran/TSG_GERAN/GERAN_65_Shanghai/Docs/GP-150001.zip. [Accessed: September 13, 2020]

[60] Ngu AHH, Gutierrez M, Metsis V, Nepal S, Sheng MZ. IoT middleware: A survey on issues and enabling technologies. IEEE Internet of Things Journal. 2016;**4**:1-20. DOI: 10.1109/jiot.2016.2615180

[61] Yuriyama M, Kushida T. Sensor-cloud infrastructure—physical sensor management with virtualized sensors on cloud computing. In: Proceedings of the International Conference on Network-Based Information Systems (N-BIS); 14-16 September 2010; Japan. New York: IEEE; 2010. pp. 1-8

[62] Idrissi ME, Elbeqqali O, Riffi J. From cloud computing to fog computing: Two technologies to serve IoT—A review-. In: Proceedings of the IEEE International Smart Cities Conference (ISC2); 14-17 October 2019; Morocco. New York: IEEE; 2020. pp. 272-279

[63] Sarkar S, Chatterjee S, Misra S. Assessment of the suitability of fog computing in the context of internet of things. IEEE Transactions on Cloud Computing. 2018;**6**:46-59. DOI: 10.1109/tcc.2015.2485206

[64] Pfandzelter T, Bermbach D. IoT data processing in the fog: Functions, streams, or batch processing? In: Proceedings of the IEEE International Conference on Fog Computing (ICFC); 24-26 June 2019; Czech Republic. New York: IEEE; 2019. pp. 201-206

[65] Yannuzzi M, Milito R, Serral-Gracia R, Montero D, Nemirovsky M. Key ingredients in an IoT recipe: Fog computing, cloud computing, and more fog computing. In: Proceedings of the IEEE International Workshop on Computer Aided Modeling and Design of Communication Links and Networks (CAMAD); 1-3 December 2014; Greece. New York: IEEE; 2015. pp. 325-329

[66] La QD, Ngo MV, Dinh TQ, Quek TQS, Shin H. Enabling

intelligence in fog computing to achieve energy and latency reduction. Digital Communications and Networks. 2019;**5**:3-9. DOI: 10.1016/j.dcan.2018.10.008

[67] Yi S, Hao Z, Qin Z, Li Q. Fog computing: Platform and applications. In: Proceedings of the IEEE Workshop on Hot Topics in Web Systems and Technologies (HotWeb); 12-13 November 2015; USA. New York: IEEE; 2016. pp. 73-78

[68] Torky M, Hassanein AE. Integrating blockchain and the internet of things in precision agriculture: Analysis, opportunities, and challenges. Computers and Electronics in Agriculture. 2020;**178**:105476. DOI: 10.1016/j.compag.2020.105476

[69] Zheng Z, Xie S, Dai H, Chen X, Wang H. An overview of blockchain technology: Architecture, consensus, and future trends. In: Proceedings of the IEEE International Congress on Big Data; 25-30 June 2017; USA. New York: IEEE; 2017. pp. 557-564

[70] Report Linker. Blockchain in Agriculture Market, Application, Provider, Organization Size and Region—Global Forecast to 2023 [Internet]. 2020. Available from: https://www.reportlinker.com/p05581101/Blockchain-in-Agriculture-Market-ApplicationProvider-Organization-Size-And-Region-Global-Forecast-to.html [Accessed: February 24, 2021]

[71] Lan G, Christopher B, Jacco S, Anton S, Jan T, van Frans D, et al. Blockchain for agriculture and food: Findings from the pilot study (2017). PB - Wageningen Economic Research. :2017. Available from: https://edepot.wur.nl/426747. [Accessed: March 24, 2021]

[72] Friha O, Ferrag MA, Shu L, Nafa M. A robust security framework based on Blockchain and SDN for fog computing enabled agricultural internet of things. In: Proceedings of the International Conference on Internet of Things and Intelligent Applications (ITIA); 27-29 November 2020; China. New York: IEEE; 2021. pp. 1-5

[73] Ferrag MA, Shu L, Yang X, Derhab A, Maglaras L. Security and privacy for Green IoT-based agriculture: Review, blockchain solutions, and challenges. IEEE Access. 2020;**8**:32031-32053. DOI: 10.1109/access.2020.2973178

[74] Jiang W, Li H, Xu G, Wen M, Dong G, Lin X. PTAS: Privacypreserving thin-client authentication scheme in blockchain-based PKI. Future Generation Computer Systems. 2019;**96**:185-195. DOI: 10.1016/j.future.2019.01.026

[75] Shen M, Tang X, Zhu L, Du X, Guizani M. Privacy-preserving support vector machine training over blockchain-based encrypted IoT data in smart cities. IEEE Internet of Things Journal. 2019;**6**:7702-7712. DOI: 10.1109/jiot.2019.2901840

[76] Novo O. Blockchain meets IoT: An architecture for scalable access management in IoT. IEEE Internet of Things Journal. 2018;**5**:1184-1195. DOI: 10.1109/jiot.2018.2812239

[77] Ding S, Cao J, Li C, Fan K, Li H. A novel attribute-based access control scheme using blockchain for IoT. IEEE Access. 2019;**7**:38431-38441. DOI: 10.1109/access.2019.2905846

[78] Dedeoglu V, Jurdak R, Putra GD, Dorri A, Kanhere SS. A trust architecture for blockchain in IoT. arXiv 2019. Preprint arXiv:1906.11461

[79] Si H, Sun C, Li Y, Qiao H, Shi L. IoT information sharing security mechanism based on blockchain technology.

Future Generation Computer Systems. 2019;**101**:1028-1040. DOI: 10.1016/j.future.2019.07.036

[80] Hammi MT, Hammi B, Bellot P, Serhrouchni A. Bubbles of trust: A decentralized blockchain-based authentication system for IoT. Computers & Security. 2018;**78**:126-142. DOI: 10.1016/j.cose.2018.06.004

[81] Paul S, Joy JI, Sarker S, Shakib AAH, Ahmed S, Das AK. An unorthodox way of farming without intermediaries through Blockchain. In: Proceedings of the International Conference on Sustainable Technologies for Industry 4.0 (STI); 24-25 December 2019; Bangladesh. New York: IEEE; 2020. pp. 1-6

[82] Abraham A, Santosh Kumar MB. A study on using private-permissioned blockchain for securely sharing farmers data. In: Proceedings of the Advanced Computing and Communication Technologies for High Performance Applications (ACCTHPA); 2-4 July 2020; India. New York: IEEE; 2020. pp. 103-106

[83] Ziechmann K. Intro to Ethereum [Internet]. 2021. Available from: https://ethereum.org/en/developers/docs/intro-to-ethereum/. [Accessed: April 02, 2021]

[84] Umamaheswari S, Sreeram S, Kritika N, Jyothi Prasanth DR. BIoT: Blockchain based IoT for agriculture. In: Proceedings of the International Conference on Advanced Computing (ICoAC); 18-20 December 2019; India. New York: IEEE; 2020. pp. 324-327

[85] Chun-Ting P, Meng-Ju L, Nen-Fu H, Jhong-Ting L, Jia-Jung S. Agriculture Blockchain service platform for farm-to-fork traceability with IoT sensors. In: Proceedings of the International Conference on Information Networking (ICOIN); 7-10 January 2020; Spain. New York: IEEE; 2020. pp. 158-163

[86] Ur Rahman M, Baiardi F, Ricci L. Blockchain smart contract for scalable data sharing in IoT: A case study of smart agriculture. In: Proceedings of the Global Conference on Artificial Intelligence and Internet of Things (GCAIoT); 12-16 December 2020; United Arab Emirates. New York: IEEE; 2021. pp. 1-7

[87] Dasaklis T, Casino F. Improving vendor-managed inventory strategy based on Internet of Things (IoT) applications and Blockchain technology. In: Proceedings of the IEEE International Conference on Blockchain and Cryptocurrency (ICBC); 14-17 May 2019; Korea (South). New York: IEEE; 2019. pp. 50-55

[88] Casino F, Dasaklis TK, Patsakis C. Enhanced vendor-managed inventory through Blockchain. In: Proceedings of the South-East Europe Design Automation, Computer Engineering, Computer Networks and Social Media Conference (SEEDA-CECNSM); 20-22 September 2019; Greece. New York: IEEE; 2019. pp. 1-8

[89] Nguyen DH, Tuong NH, Pham HA. Blockchain-based farming activities tracker for enhancing trust in the community supported agriculture model. In: Proceedings of the International Conference on Information and Communication Technology Convergence (ICTC); 21-23 October 2020; Korea (South). New York: IEEE; 2020. pp. 737-740

[90] Yang X, Li M, Yu H, Wang M, Xu D, Sun C. A trusted blockchain-based traceability system for fruit and vegetable agricultural products. IEEE Access. 2021;**9**:36282-36293. DOI: 10.1109/ACCESS.2021.3062845

[91] Madumidha S, Ranjani PS, Vandhana U, Venmuhilan B. A theoretical implementation: Agriculture-food supply

chain management using Blockchain technology. In: Proceedings of the TEQIP III Sponsored International Conference on Microwave Integrated Circuits, Photonics and Wireless Networks (IMICPW); 22-24 May 2019; India. New York: IEEE; 2019. pp. 174-178

[92] Grand view research. Farm Management Software Market Size Worth $4.22 Billion By 2025 [Internet]. 2018. Available from: https://www.grandviewresearch.com/press-release/global-farm-management-software-market. [Accessed: April 01, 2021]

[93] Angin P, Anisi MH, Göksel F, Gürsoy C, Büyükgülcü A. AgriLoRa: A digital twin framework for smart agriculture. Journal of Wireless Mobile Networks, Ubiquitous Computing, and Dependable Applications. 2020;**11**:77-96. DOI: 10.22667/JOWUA.2020.12.31.77

[94] Weltzien C, Shamshiri RR. SunBot: Autonomous nursing assistant for emission-free berry production, general concepts and framework. In: VDI Wissensforum GmbH, editors. LAND.TECHNIK AgEng 2019: The Forum for Agricultural Engineering Innovations. 2019 ed. Düsseldorf: VDI Verlag; 2019. pp. 463-470. DOI: 10.51202/9783181023617-463

[95] Shamshiri RR, Hameed IA, Balasundram SK, Ahmad D, Weltzien C, Yamin M. Fundamental research on unmanned aerial vehicles to support precision agriculture in oil palm plantations. In: Zhou J, Zhang B, editors. Agricultural Robots-Fundamentals and Application. 2018th ed. Rijeka: IntechOpen; 2019. DOI: 10.5772/intechopen.80936

[96] Yamin M, Ismail WIW, Kassim MSM, Aziz SBA, Akbar FN, Shamshiri RR, et al. Modification of colorimetric method based digital soil test kit for determination of macronutrients in oil palm plantation. International Journal of Agricultural and Biological Engineering. 2020;**13**:188-197. DOI: 10.25165/j.ijabe.20201304.5694

[97] Chlingaryan A, Sukkarieh S, Whelan B. Machine learning approaches for crop yield prediction and nitrogen status estimation in precision agriculture: A review. Computers and Electronics in Agriculture. 2018;**151**:61-69. DOI: 10.1016/j.compag.2018.05.012

[98] Klompenburg VT, Kassahun A, Catal C. Crop yield prediction using machine learning: A systematic literature review. Computers and Electronics in Agriculture. 2020;**177**:105709. DOI: 10.1016/j.compag.2020.105709

[99] Nevavuori P, Narra N, Lipping T. Crop yield prediction with deep convolutional neural networks. Computers and Electronics in Agriculture. 2019;**163**:104859. DOI: 10.1016/j.compag.2019.104859

[100] Shamshiri RR, Bojic I, van Henten E, Balasundram SK, Dworak V, Sultan M, et al. Model-based evaluation of greenhouse microclimate using IoT-sensor data fusion for energy efficient crop production. Journal of Cleaner Production. 2020;**263**:121303. DOI: 10.1016/j.jclepro.2020.121303

[101] Li M, Shamshiri RR, Schirrmann M, Weltzien C. Impact of camera viewing angle for estimating leaf parameters of wheat plants from 3D point clouds. Agriculture. 2021;**11**:563. DOI: 10.3390/agriculture11060563

[102] Shamshiri RR, Weltzien C. Development and field evaluation of a multichannel LoRa sensor for IoT monitoring in berry orchards. In: Meyer-Aurich A, Gandorfer M, Hoffmann C, Weltzien C,

Bellingrath-Kimura S, Floto H, editors. 41. GIL-Jahrestagung, Informations- und Kommunikationstechnologie in kritischen Zeiten. Bonn: Gesellschaft für Informatik e.V. pp. 289-294

[103] Shamshiri R, Ehsani R, Maja MJ, Roka MF. Determining machine efficiency parameters for a citrus canopy shaker using yield monitor data. Applied Engineering in Agriculture. 2013;**29**:33-41. DOI: 10.13031/2013.42526

[104] Shamshiri RR, Weltzien C, Hameed IA, Yule IJ, Grift TE, Balasundram SK, et al. Research and development in agricultural robotics: A perspective of digital farming. International Journal of Agricultural and Biological Engineering. 2018;**11**:1-14. DOI: 10.25165/j.ijabe.20181104.4278

[105] Shamshiri RR, Hameed IA, Karkee M, Weltzien C. Robotic harvesting of fruiting vegetables: A simulation approach in V-REP, ROS and MATLAB. In: Hussmann S, editor. Automation in Agriculture—Securing Food Supplies for Future Generations. 2018th ed. Rijeka: IntechOpen; 2018. DOI: 10.5772/intechopen.73861

[106] Lee J, Kim Y. Preventing bitcoin selfish mining using transaction creation time. In: Proceedings of the International Conference on Software Security and Assurance (ICSSA); 26-27 July 2018; Korea (South). New York: IEEE; 2020. pp. 19-24

[107] IEEE. IEEE future directions blockchain white paper. Reinforcing the Links of the Blockchain [Internet]. 2017. Available from: https://blockchain.ieee.org/images/files/pdf/ieee-future-directions-blockchain-white-paper.pdf. [Accessed: March 20, 2021]

[108] Thaiyalnayaki K, Joseph C. Classification of plant disease using SVM and deep learning. Materials Today: Proceedings. 2021;**47**:468-470. DOI: 10.1016/j.matpr.2021.05.029

[109] Atila Ü, Uçar M, Akyol K, Uçar E. Plant leaf disease classification using EfficientNet deep learning model. Ecological Informatics. 2021;**61**:101182. DOI: 10.1016/j.ecoinf.2020.101182

[110] Chen M, Cui Y, Wang X, Xie H, Liu F, Luo T, et al. A reinforcement learning approach to irrigation decision-making for rice using weather forecasts. Agricultural Water Management. 2021;**250**:106838. DOI: 10.1016/j.agwat.2021.106838

[111] Ju S, Lim H, Ma JW, Kim S, Lee K, Zhao S, et al. Optimal county-level crop yield prediction using MODIS-based variables and weather data: A comparative study on machine learning models. Agricultural and Forest Meteorology. 2021;**307**:108530. DOI: 10.1016/j.agrformet.2021.108530

[112] Guerrero JM, Ruz JJ, Pajares G. Crop rows and weeds detection in maize fields applying a computer vision system based on geometry. Computers and Electronics in Agriculture. 2017;**142**:461-472. DOI: 10.1016/j.compag.2017.09.028

[113] Grijpink F, Kutcher E, Ménard A, Ramaswamy S, Schiavotto D, Manyika J, et al. Connected world An evolution in connectivity beyond the 5G revolution [Internet]. 2020. Available from: https://www.mckinsey.com/~/media/mckinsey/industries/technology%20media%20and%20telecommunications/telecommunications/our%20insights/connected%20world%20an%20evolution%20in%20connectivity%20beyond%20the%205g%20revolution/mgi_connected-world_discussion-paper_february-2020.pdf. [Accessed: January 20, 2021]

[114] Jawad H, Nordin R, Gharghan S, Jawad A, Ismail M. Energy-efficient

wireless sensor networks for precision agriculture: A review. Sensors. 2017;**17**:1781. DOI: 10.3390/s17081781

[115] Villa-Henriksen A, Edwards GTC, Pesonen LA, Green O, Grøn Sørensen CA. Internet of things in arable farming: Implementation, applications, challenges and potential. Biosystems Engineering. 2020;**191**:60-84. DOI: 10.1016/j.biosystemseng.2019.12.013

[116] Bauer J, Aschenbruck N. Design and implementation of an agricultural monitoring system for smart farming. In: Proceedings of the IEEE IoT Vertical and Topical Summit on agriculture (IOT-Tuscany); 8-9 May 2018; Italy. New York: IEEE; 2018. pp. 1-6

[117] PwC Global. PwC Global Blockchain Survey_Executive Summary_HK_v4 [Internet]. 2018. Available from: https://www.pwccn.com/en/research-and-insights/publications/global-blockchain-survey-2018/global-blockchain-survey-2018-survey-highlights.pdf. [Accessed: March 24, 2021]

Chapter 2

Perspective Chapter: Perspectives on Pathogenic Plant Virus Control with Essential Oils for Sustainability of Agriculture 4.0

Thanat Na Phatthalung and Wipa Tangkananond

Abstract

The outbreaks of plant pathogenic viruses and insect pests affect agricultural product supply chain systems. Environmentally friendly innovative technologies are provided accurate, practical, and acceptable means for surveillance by farmers. The bioactive compound applications are derived from plant essential oils with antiviral activities as well as integrating insect pest control and management are useful choices. Successful comprehensive planning, including material production systems, extraction techniques, quality testing, and product creation are essential for strategic and operational decision-making under current operation management trends of Agriculture 4.0. This information can potentially be used to impel today agriculture and set the directions for supports. The role of management and data analysis will meet the challenges of increasing populations and food security with the ultimate goal to achieve efficient and sustainable effectiveness for all participants in directing the world agricultural systems.

Keywords: plant virus, plant essential oils, biopesticides, innovative technology, agriculture 4.0

1. Introduction

The world population has been increasing continuously that is anticipated to reach about 9.7 billion by 2050 and predicted to be 11.2 billion by 2100 [1]. This will be an important factor for the directional determination in agricultural management, which impacted the human population, environment, and ecosystems. These challenges should be systematically managed by integrating with the environmentally friendly innovative technology of Agricultural 4.0. The pests and plant diseases management agents will be based on natural products or biopesticides are the great promise in controlling yield quality. However, this agricultural management with natural products could be taken continually in steps to boost the consumption in the global market, which will likely increase in the future for replacing and reducing the chemical pesticides use. Presently, plant essential oil (plant EO) derived biopesticides are assessed and accepted in many countries through the public or specific regulation uses for assessing the active compounds and substances. The suitable extraction methods are supported to create the natural product, which are the important operations of determining the biological activities of plant EO.

Therefore, this chapter will describe the application of plant EOs for antiviral activities and insect-pest managements as well as discussing the relative innovative technologies such as automation, smart devices, smart sensors, artificial intelligence (AI), novel techniques and technologies, and the Internet of things (IoT). These will be applied under sustainable agricultural managements by Agriculture 4.0. These integrated operations in various innovative technologies will be made it possible to be quickly successful to increase the role of natural products in sustainable agricultural managements.

2. Perspectives on the potential of plant essential oils as green biopesticides in agricultural 4.0

Plant essential oils (Plant Eos) were used as biopesticides in agricultural systems for a long time. In the case of local usage, the plant materials were extracted by using differently traditional extraction techniques that the quantity and quality of bioactive essential oil compounds (EOCs) were less [2]. Therefore, the local knowledge will be upgraded for commercial production. Natural products will be continually accepted and used by farmers in the epidemic areas. A competitive challenge for commercial producers has high competition and follows by the trends of environment and healthy consumption under the world market. The environmental contamination and human health problems caused by the overuse of chemical pesticides have been reported and published in recent years. The use of chemical pesticides was the first choice for pest management, which has been increasingly apparent because of the high efficiency, specificity, and fast-acting on the target insect vectors [3].

The phenomenon, which related to the increasing use of chemical pesticides in agriculture, was the result of the successful breeding of new high-yielding rice varieties in the green revolution period [4]. The various innovative agricultural technological achievements over the years of synthetic chemical products were shown a fast action and specific effect on target organisms but have developed resistance against them. Thus, this awareness regarding problems had been significantly important to the agricultural management. Especially in research development pertained to avoid chemical resistance of insect pests by the green innovative technology and integrating sustainability principles [5]. In this context, biopesticides derived from the different plant species have the potential for solving problems as well as developing natural commercial products for safe crop productivity increasing. Biopesticides are becoming a bright alternative replacement to chemical pesticides due to the significantly growing agricultural supply chain of both consumers and producers. However, there are limitations of plant EO activities such as rapid conversion and degradation by the various factors under field conditions.

Nowadays, despite considerable research and development effort on the plant EO properties and their active compound, yet their commercial products have few appeared in the global marketplace. As a result of this, it cannot be denied that such issues are only achieved concrete results at the policy level, which resulted from the regulatory commercialization barriers. Therefore, the status and potential of plant EOs as green biopesticides should be researched and developed with innovative technology under the three concepts including social acceptability,

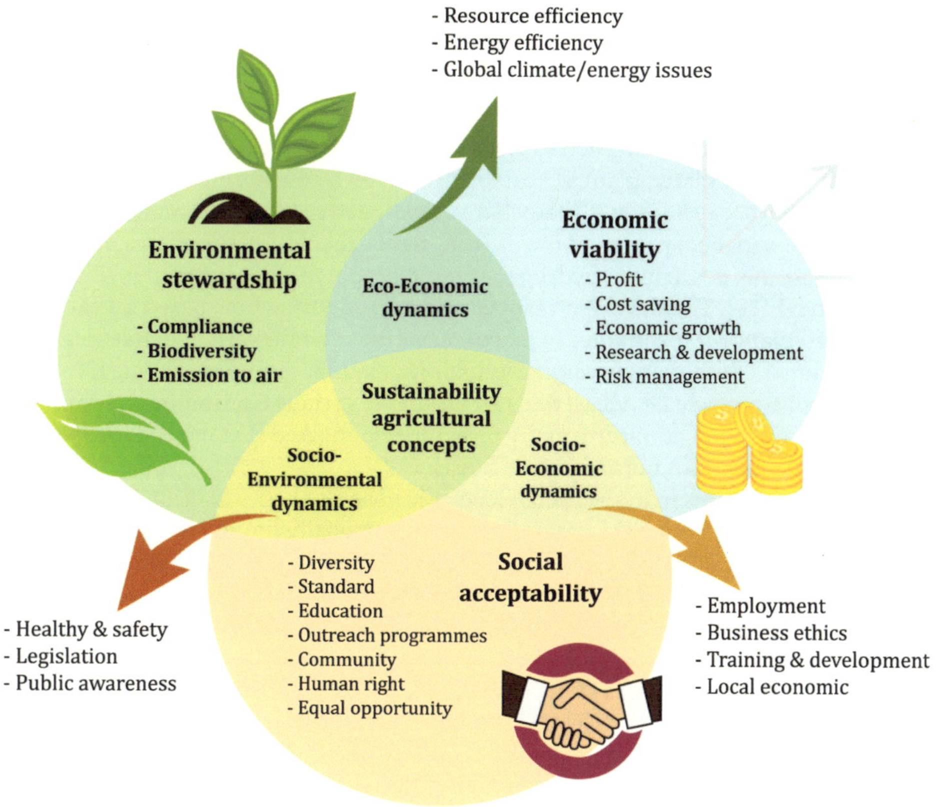

Figure 1.
The sustainability agricultural management concepts. (figure was created from reference number [6]).

economic viability, and environmental stewardship (**Figure 1**) [6]. As a result, the high-quality products will be created with low cost and easy to use in the operation model of sustainability Agriculture 4.0.

3. Innovative technologies of plant essential oil extraction and quality control

The conventional extraction methods were heated for a long extraction time, and they depended on extracting solvents from various extraction procedures such as maceration (MA), soxhlet extraction (SE) [7], sonication/ultrasonication extraction (USE) [8], steam distillation (SD) [9], and solid–liquid extraction (SLE) [10]. The bioactive EOCs were destroyed, concentration reduced, lowered down reproducibility and extraction efficiency. These methods had used large content of plant materials and organic solvents, which were the main inefficiencies of natural resource use. The innovative technologies are environmentally friendly for plant EO extraction, constantly being invented and developed for efficient use of various resources. Using high-efficiency and uncomplicated extraction techniques will reduce the production costs of natural resources such as pressurized liquid extraction (PLE) [11], supercritical fluid extraction (SFE) [12], ultrasound-assisted extraction (UAE) [13], microwave-assisted extraction (MAE) [14], pulsed electric field extraction (PEFE) [15], enzyme assisted extraction (EAE) [16], solvent-free microwave extraction (SFME), and headspace solid-phase microextraction (HS-SPME). They also increase the yield of the bioactive compounds with the high quality of extract.

Application usages of these innovative extraction technologies are interesting alternative ways for enhancing active plant EO properties and efficiencies. The stability and quantity of isolated plant EO can be preserved by encapsulation forms (e.g., droplets, particles, capsules, multilamellar vesicles, active film, and complexes) [17] and polymeric nanoencapsulation forms (e.g., nanocapsules, nanospheres, miscelle, nanogel, liposome, dendrimer, hydrogel, layered biopolymer, mesoporous silica, and nanofiber) [18]. The developed biopesticides products, which based on various encapsulated plant EO techniques (e.g., coacervation, complexation, emulsification, film hydration method, nanoprecipitation, ionic gelation, and spray drying), can slowly and continuously be released to targets under various environmental conditions. According to the literature, many researchers reported that nano-active forms had more efficiency than normal-active forms.

Interesting advances in innovation, electronic nose (E-nose, EN) techniques can be applied for quality control of natural products, especially the volatile organic compounds (VOCs) [19]. The biological olfactory detector system called E-nose sensor technique is based on different electronic aroma detection (EAD) technologies by gas sensors. These are as follows: bulk acoustic wave (BAW), surface acoustic wave (SAW), calorimetric/catalytic bead (CB), carbon black composite (CBC), conductive polymers (CP), electrochemical sensors (EC), fluorescence (FL), metal-oxide semiconductors (MOS), complementary MOS (CMOS), MOS field-effect transistors (MOSFET), micro-electromechanical systems (MEMS), optical fiber live cell (OF-LC), and quartz crystal microbalance (QCM) [20–23]. In addition, E-nose instrument consists of both hardware and software components [24]. They include (1) sensors and chemicals that the specific sensors are designed to convert the chemical information of VOCs into analytical signals; (2) machine learning (ML) algorithms act an information-processing unit such as linear discriminant analysis (LDA), quadratic discriminant analysis (QDA), discriminant function analysis (DFA), stepwise discriminant analysis (SDA), partial least squares regression (PLSR), generalized least squares regression (GLSR), multiple linear regression (MLR), principle component analysis (PCA), support vector machines (SVMs), k-nearest neighbor analysis (KNN), artificial neural networks (ANNs), and genetic algorithms (GA) [25–27], and all pattern-recognition algorithms were processed: data collection, modeling, training, and evaluation; and (3) system performance evaluation, which the results have been calculated through E-nose system evaluation metrics with accuracy, precision, sensitivity, specificity, and F1-score (harmonic mean). These were incorporated with reference-library databases [28] with (4) both sensor types and application of commercially available E-noses.

Applications of E-nose technologies for the development and monitoring control of plant EOs were performed and operated in industrial processes. Rasekh et al. [28] and Rasekh et al. [29], for instance, showed that the developed method of E-nose systems with nine MOSs (MAU-9 MOS E-nose system), and two statistical analyses of LDA and QDA methods were successfully evaluated for quickly identifying and classifying plant EOs derived from fruit and herbal edible-plant sources. The developed E-nose array with statistical methods was shown the discrimination results into two groups of fruits and herbal plant EO types with 100% correct accuracy in both LDA and QDA methods and the classification results of different plant EO sample types with the correct accuracy of LDA (98.9%) and QDA (100%), including tarragon oil (*Artemisia dracunculus* L., Asteraceae), thyme oil (*Thymus vulgaris* L., Lamiaceae), cornmint oil (*Mentha arvensis* L., Lamiaceae), lemon oil (*Citrus limon* L. Burm. f., Rutaceae), orange oil (*C. sinensis* L., Rutaceae), and mango oil (*Mangifera indica* L., Anacardiaceae). Similarly, Okur et al. [30] identified the different six species of mints (family Lamiaceae) by the QCM sensors and digital pattern-recognition algorithms of PCA, LDA, and KNN. The mint species were classified accurately by the statistical methods of PCA (97.2%), LDA (100%), and KNN (99.9%) and include peppermint (*M. piperita* L.), spearmint (*M. spicata* L.), curly mint (*M. spicata* ssp. crispa), horsemint (*M. longifolia* L.), Korean mint [*Agastache rugosa* (Fisch. & C.A.Mey.) Kuntzeand], and catmint (*Nepeta cataria*. L.). Similar results have been reported in the various plant VOCs of edible plant species [31], tomato [32], and apple [33]. Graboski et al. [34] reported that the developable method of carbon nanocomposites (CNC) E-nose system was capable to detect the distinction between the plant EO of clove [*Syzygium aromaticum* (L.) Merr. & L.M.Perry, Myrtaceae], eugenol, and eugenyl acetate. Moreover, Lias et al. [35] found that the E-nose system depicted a strong correlation between sample volume and sensors intensity values to plant EO composition of agarwood. In another study, Wu et al. [36] demonstrated that an ultra-fast gas chromatography (UFGC)-type E-nose system was identified the VOCs of spikenard (*Nardostachys chinensis* Batalin, Valerianaceae) with 94% accuracy. Significantly, the E-nose systems and digital pattern-recognition algorithms were used to classify different plant species and varieties such as garlic (*Allium* spp.) [37], pepper (*Capsicum* spp.) [38], and cucumber [39]. Based on the literature review, E-nose technologies and digital pattern-recognition algorithms are potential and effective safety tools for the rapid detection, identification, verification, and validation of plant EOs of plant materials and commercial plant products as environmentally friendly biopesticides in the strategy and policy of sustainable agricultural management.

4. Antiviral activity mechanisms and their applications

The application of plant EOs and active components as direct or indirect effects of antiviral or virucidal activity together with the insect pest control and management [40] is an interesting operation. Many research studies have been focused on medicinal pathogenic human and animal viruses. This knowledge can be further database documented, developed, and applied to plant pathogenic viruses and insect vectors for data-driven agriculture and management.

The plant EOs and their components have been effective in increasing physical/chemical/biological stabilities and their antiviral effectiveness. Several research studies were reported the potential plant EOs for antiviral activity, for instance, showed that the plant EO isolated from star anise (*Illicium verum* Hook.f., Illiciaceae) and fennel (*Foeniculum vulgare* Mill., Apiacae) had potentially inhibited *Potato virus X* (PVX: *Potexvirus*, *Flexiviridae*), *Tobacco ringspot virus* (TRSV: *Nepovirus*, *Secoviridae*), and *Tobacco mosaic virus* (TMV: *Tobamovirus*, *Virgaviridae*). Similarly, Bishop [41] found that the local lesions of TMV on tobacco (*Nicotiana glutinosa* L., Solanaceae) decreased after being tested by the tea tree oil [*Melaleuca alternifolia* (Maiden & Betche) Cheel., Myrtaceae]. In relation, Iftikhar et al. [42] tested the EO of clove [*S. aromaticum* (L.) Merr. & L.M.Perry, Myrtaceae] caused maximum inhibition of *Potato leaf roll virus* (PLRV: *Polerovirus*, *Luteoviridae*). Lu et al. [43] reported that TMV transmission was inhibited by the EO of artemisia (*Artemisia vulgaris* L., Asteraceae), ginger (*Zingiber officinale* Roscoe, Zingiberaceae), and lemongrass [*Cymbopogon citratus* (Dc. Ex Nees) Stapf, Gramineae]. Moreover, Dikova et al. [44] found that lavender oil (*Lavandula angustifolia* Mill., Lamiaceae) could control *Tomato spotted wilt virus* (TSWV: *Tospovirus*, *Bunyaviridae*). The EOs extracted from billygoat-weed (*Ageratum conyzoides* L., Asteraceae), bottle brush [*Callistemon citrinus* (Curtis) Skeels, Myrtaceae], ajwain (*Carum copticum* L., Apiaceae), holy basil (*Ocimum sanctum* L., Lamiaceae), and pepper elder [*Peperomia pellucida* (L.) Kunth, Piperaceae] have potentially inhibited *Cowpea mosaic virus* (CPMV: *Comovirus*, *Comoviridae*), *Bean common mosaic virus* (BCMV: *Potyvirus*, *Potyviridae*), and *Southern bean mosaic virus* (SBMV: *Sobemovirus*, *Solemoviridae*) [45]. In another study, Helal [46] reported that the plant EOs of thyme (*T. vulgaris* L., Lamiaceae) and peppermint (*M. piperita* L., Lamiaceae) had inhibition effects of *Tobacco necrosis virus* (TNV: *Necrovirus*, *Tombusviridae*) and *Cucumber mosaic virus* (CMV: *Cucumovirus*, *Bromoviridae*).

According to recent studies, Na Phatthalung and Tangkananond [47] applied dot-immunobinding assay (DIBA) for evaluating the potential of plant EO for transmission inhibitory effects on *Rice ragged stunt virus* (RRSV: *Oryzavirus*, *Reoviridae*) by the brown planthopper (BPH: *Nilaparvata lugens* Stål) (Homoptera: Delphacidae). These studies were demonstrated that all the tested plant EO had potential transmission inhibitory in efficiency ranges from 0.002 to 0.1% from the infected rice plants to non-viruliferous BPH status. In addition, viruliferous BPH status was communicated with similar success to viral-free rice plants. These include black pepper (*Piper nigrum* L., Piperaceae), lemongrass, star anise, kaffir lime (*Citrus hystrix* DC, Rutaceae), and kaempfer [*Boesenbergia rotunda* (L.) Mansf., Zingiberaceae] highly effected 10–70% inhibition and lime [*C. aurantifolia* (Christm.) Swingle, Rutaceae], galangal [*Alpinia galangal* (L.) Sw., Zingiberaceae], holy basil, sweet basil (*O. basilicum* L., Lamiaceae), and betelvine (*P. betle* L., Piperaceae) slightly effected 10–30% inhibition, respectively (**Figure 2**). Furthermore, the plant EOs of star anise and lemongrass were selected for assessing the toxicity and physiological effects on the BPH vector. These results showed that the plant EO in the range from 3 to 5% showed malformed structures and completely destroyed within 3–5 days after treatment (DAT) (**Figure 3**). Therefore, the plant EOs paved the possibility and potential candidates for further prototype development as commercial antiviral agents for plant protection and sustainable agricultural management in agriculture 4.0.

Figure 2.
The potential of plant EO for transmission inhibitory effects on RRSV by the BPH vector and detection method by DIBA (figure was modified from reference number [47]).

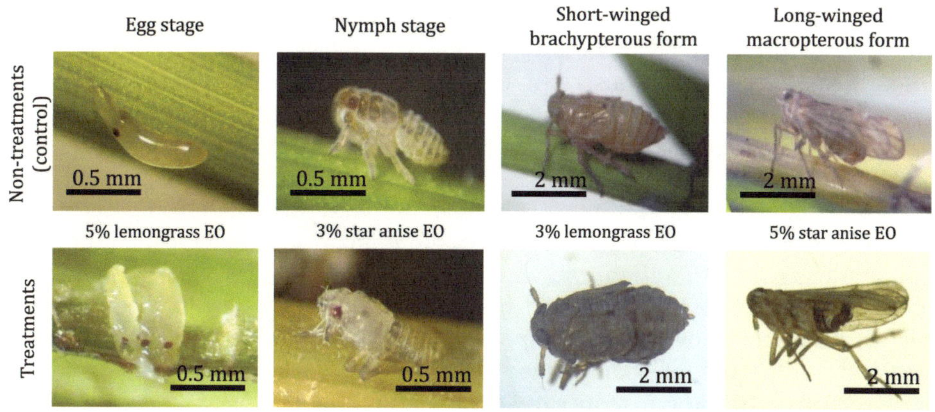

Figure 3.
The morphological effects of plant EOs on the BPH. (figure was modified from reference number [47]).

It is possible to hypothesize the antiviral mechanisms from the literature reviews about the viral infection cycle in host cell-culture-based systems (*in vitro*) and viral host models (*in vivo*) as well as molecular docking (*in silico*) [48–50]. The summary concept of antiviral mechanisms by plant EO can be divided into direct and indirect actions. Several modes of direct antiviral actions affected the enveloped and non-enveloped (naked) viral progenies by substance and enzyme blocking in different steps of the viral infection cycle (**Figure 4**) [51, 52]. The various plant EOs and active components have potential inactivation viral activities, transmissibility, stability, and infectivity on enveloped viruses more than on the naked viruses [51].

Several modes of indirect antiviral actions affect host properties, viral transmission modes, and infection efficiency. Generally, plant EO has important features

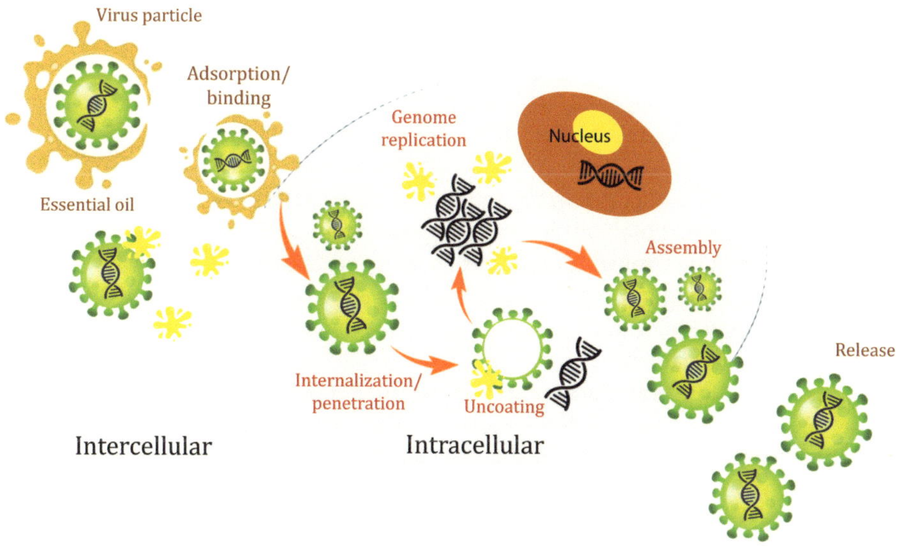

Figure 4.
The mechanism of antiviral actions as possible targets for plant EO. (figure was modified from reference number [51]).

of hydrophobic properties including surface tension, contact angle value, droplet volumes, and lubricating with varied viscosities that affect the external surface area structure properties of viral hosts [53]. Insect vectors or plants that were sprayed with plant EO may be modified the physiology and disturb the metabolism of the inoculated cells [54]. External surface areas of the viral particles and hosts were coated, which affect the infectivity and transmissibility, were inhibited. Developmental and survival periods of insect vectors are significant for viral transmission and nymph stages are most important for viral transmission. Adult stages are important for population increase, migration, and viral spread [55]. However, several plant EOs tended to be more effective on the soft-bodied insects than the hard-bodied insects. They affected host plant manipulation by the induced systemic resistance (ISR) and insecticidal properties [56]. The active plant EO can manage the insect vector damaging effects on crops and also reduce their plant viral transmission ability.

The plant EO has optimal properties for covering with the general surface structure of probing stylet or body-cuticle (extracellular layer) of insect vectors and has optimal activities for viral transmission inhibition. Therefore, the inhibition of virus transmission by plant EO occurs at the virus-vector or virus-vector-plant relationships (tri-partite relationships). All of these significantly play an important role for knowledge applying in future crop protection and successful pest management under the Agriculture 4.0 policy.

5. Current status progress of plant EOs and active compounds for sustainability in agriculture 4.0

Using the status of applied plant EOs has not seen any concrete results in the continued practical use of farmers. Farmer occupation is mainly for life subsistence as well as lack of business processes in response to the policy of Agriculture 4.0. Therefore, the use of plant EO will be part of the chain of production processes until the plantation level to prepare the quality of raw materials. Additionally, active network information should be published to build the acceptance and confidence with the integration of agricultural knowledge, science, and technology together with the modern innovation. Network creation of a collaboration between researchers, entrepreneurs, and farmers in response to the development of intensive and comprehensive support mechanisms for agricultural innovation. The smart operating cycle based on agricultural database systems and network management organization will be helpful in efficient and comprehensive management that are shown in **Figures 5** and **6** [57–59]. Natural-product-based plant EOs can be applied for crop protection and management in the preliminary processes under farm operation. The operational results for pathogen detection rely on a more complex concept of visions as follows: data collection, processing, analysis, and publishing by smart platforms.

Sørensen et al. [60] indicated the conceptual model of a future farm management information system. Smart electronic tools with easy use and affordable prices are important factors in the real-time business decision-making for farmers under the highly competitive markets known as Farm Management Information Systems (FMIS). FMIS was integrated by various technologies and standard software packages such as information technology (IT), information systems (IS), and enterprise resource planning (ERP) in the form of information for data collection, processing, storing, and disseminating [61]. All of FMIS operations, information and multiple business functions with registration, interoperation, and communication in connection with

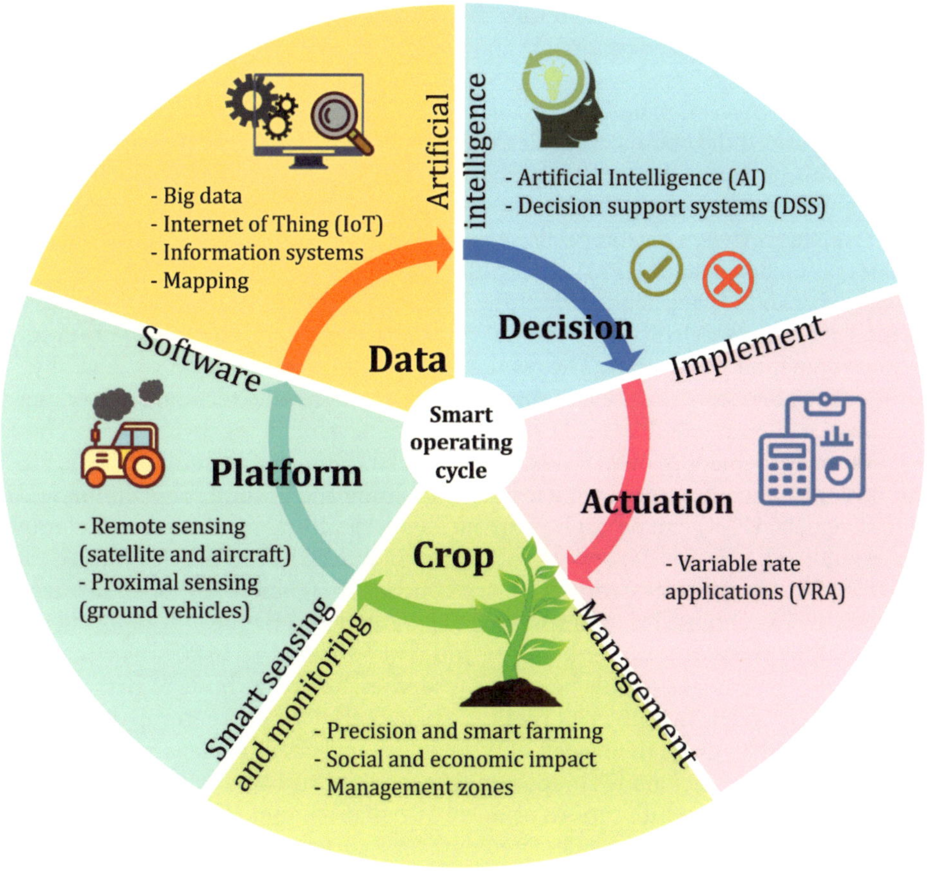

Figure 5.
The smart operating cycle based on agricultural database systems and network management organization (figure was created from figure and table of reference number [57]).

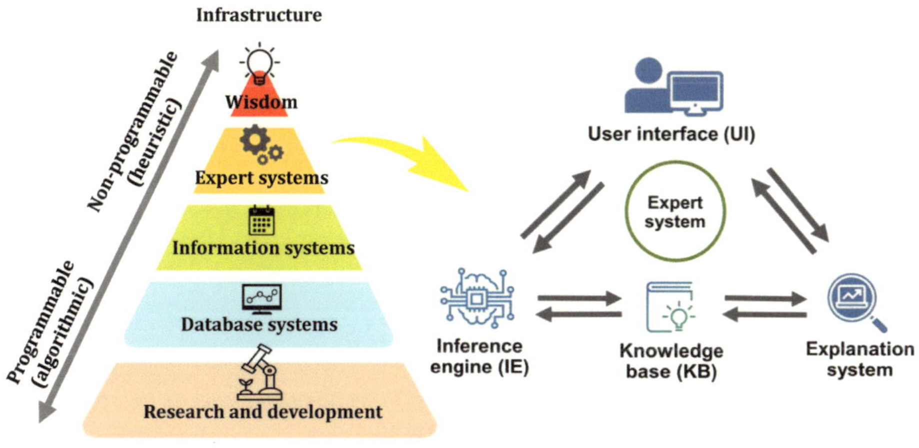

Figure 6.
The pyramid of smart agricultural operating hierarchy. (figure was modified from figures and data of reference numbers [58, 59]).

external systems were incorporated for a single integrated system creating [62]. Silvie et al. [63] showed that the developable knowledge base and a software prototype called Knomana knowledge-based system (KBS) for botanical species used as pesticide plant species for crop protection and pest management. The developable software prototype can be categorized the botanical species and their used parts for the protection of targeted organisms. It also shows the ranking of active plant species used in plant health for users and alternative information for selecting suitable methods and applications. Therefore, this software prototype also enables the novel knowledge production related to insect pest management (IPM) push-pull strategy and policy.

Pantazi et al. [64] applied the machine learning (ML) techniques connected to the internet of things (IoT) and wireless sensor network (WSN) for recognition of the environmental parameters. The results showed that this operation successfully distinguished between healthy and diseased plants. Interesting techniques, advanced technologies of automated and robotic systems are developed for precision agriculture and plant management in open fields. Plant health monitoring by remote sensing technique of unmanned aerial vehicle (UAV) or drone and ground robot (unmanned ground vehicle, UGV) can be applied for various agricultural management including crop monitoring [65], field mapping [66], plant population counting [67], weed management [68], biomass estimation [69], crop nutrient diagnosis [70], plant disease diagnosis and detection [71], and spraying [72]. Tillett and Hague [73] reported that a machine vision system could detect and remove weeds up to 80% as well as weeds could serve as susceptible hosts and reservoir alternative hosts of pathogens and their vectors. The imaging techniques have potential for various crop diseases detection including ground imagery, UAV imagery, and satellite imagery. Similarly, Mongkolchart and Ketcham [74] reported that the rice leaf color values of rice plant diseases were caused by infestations of the brown planthopper (BPH) and rice leaf-folder (RLF) and were correctly detected with 73% accuracy. Xie et al. [75] found that the application of ground imagery with deep learning (DL) methods and extreme learning machine (ELM) classifier model could detect different tobacco diseases with accuracy ranging from 97.1 to 100%. In a similar way, Zhu et al. [76] reported that the ELM classifier could be applied to the hyperspectral image (HSI) for TMV detection on tobacco leaves with 98% accuracy. In the same context, Jin et al. [77] successfully classified between infected and healthy wheat head crops by HSI with 84.6% accuracy. Therefore, the roles of image analysis in robotic management, as well as robotic systems and human-robot collaboration (co-robot) systems, have the potential for greater efficiency and flexibility in open agricultural fields and environments. These knowledge systems have a high potential for crop disease prediction and detection in earlier stages by meteorological systems integrated with algorithms. In addition, robot systems can cooperate for one-stop service development with various detection methods such as next-generation sequencing (NGS) techniques [78], loop-mediated isothermal amplification (LAMP) [79], and lab-on-chip based on electrical impedance spectroscopy (EIS) [80].

6. Biocontrol product trends and innovative technological developments for antiviral and insect-pest management

The trends of plant EO for antiviral property and insect-pest management under the sustainable agricultural crop production were not widely accepted when compared with the synthetic chemicals. The interactions of host and virus have developed resistance to bioactive compounds [51, 81]. The advantages of using natural products including; agriculture product safety, reduced levels of plant viruses and insect pests, improved product quality as well as value and guaranteed market access. However, these advantages depend on the physical factors (e.g., agro-climatic zones, seasons, and crops) and biological factors (e.g., biotransformation population dynamics of microorganisms, microbial degradation). Therefore, product development responds to a wide range of applications and is suitable for use in large-scale agricultural fields. Agriculture 4.0 policy plays an important role in the development of the preparation and processing of plant materials for the effective production of natural substances, crop protection, and successful pest management.

Several bioactive compounds of plant EOs were confirmed and classified as generally recognized as safe (GRAS) by the United States Food and Drug Administration (FDA) and the United States Environmental Protection Agency (EPA), which reported in the medical and agricultural applications. For example, thymol and carvacrol as the main compounds were isolated from winter savory (*Satureja montana* L., Lamiaceae) and showed the direct inactivation of TMV and CMV [82]. Sun et al. [83] reported that the plant-derived compound of eugenol showed effective antiviral activity of *Tomato yellow leaf curl virus* (TYLCV: *Begomovirus*, *Geminiviridae*) and induced the salicylic acid (SA) biosynthetic pathway. The main bioactive compounds of lemon-scented gum (*Eucalytpus citriodora* Hook., Myrtaceae) and fennel include eucalyptol, D-limonene, and L-limonene and eugenol in clove buds can inhibit PLRV infection [42]. Three monoterpenes (thymol, carvacrol, and p-cymene) that were extracted from charlock (*Sinapis arvensis* L., Brassicaceae), balangu (*Lallemantia royleana* Benth., Lamiaceae), and small fleabane (*Pulicaria vulgaris* Gaertn., Asteraceae) had an inhibitory effect against *Herpes simplex virus* type 1 (HSV-1: *Simplexvirus*, *Herpesviridae*) [84]. However, differences of viral types and componential diversity of plant EOs were affected the biological mechanisms in the antiviral and insecticidal activities.

Limitations of various conventional techniques for detection and analysis of bioactive compounds are separated sampling, adsorbent preference, and taking a long time. It is also requiring additional equipment such as adsorbent traps, laboratory-based molecular assays, and gas chromatography–mass spectrometry (GC–MS). While the applications of noninvasive methods and innovative technologies such as E-nose, gas chromatography–flame ionization detector (GC-FID), proton-transfer-reaction mass spectrometry (PTR-MS), proton-transfer-reaction–time of flight–mass spectrometry (PTR-TOF-MS), electrolyte-insulator–semiconductor (EIS) sensor, and image analysis systems had potential for specific compound analyses [85, 86]. The other indirect-plant disease identification methods by morphological and physiological changes can be applied in the field with smart technologies. Digital camera technologies of visible/RGB (red, green, and blue) imaging-based methods can be applied for plant phenotyping and monitoring during the growing season [74]. The hyperspectral (HS) imaging-based systems were used for TSWV detection at an early stage, which Wang et al. [87] showed successfully detected with 96.25% accuracy and the economic impact of plant viruses such as TMV [88], *Grapevine vein-clearing virus* (GVCV: *Badnavirus*, *Caulimoviridae*) [89], *Tulip breaking virus* (TBV: *Potyvirus*, *Potyviridae*) [90], and *Potato virus Y* (PVY: *Potyvirus*, *Potyviridae*) [91] similarly operated. In the same way, the alternative viral detection methods before the appearance of visible symptoms by chlorophyll fluorescence (ChlF) imaging can be potentially used for CMV [92], TMV [93], *Pepper mild mottle virus* (PMMoV: *Tobamovirus*, *Virgaviridae*) [94], *Sweet potato feathery mottle virus* (SPFMV: *Potyvirus*, *Potyviridae*), *Sweet potato chlorotic stunt virus* (SPCSV: *Crinivirus*, *Closteroviridae*) [95], and *Turnip crinkle virus*

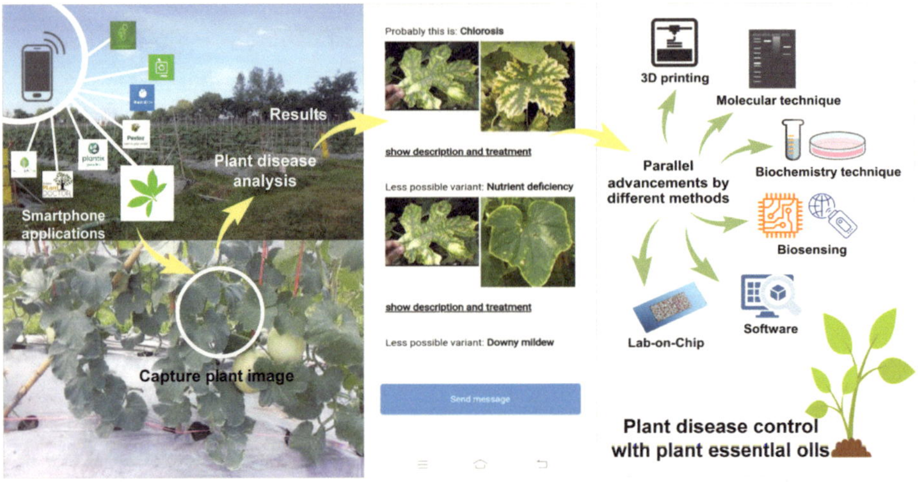

Figure 7.
Read-out platforms of smartphone applications and parallel advancements by different methods for plant disease diagnosis and detection.

(TCV: *Carmovirus*, *Tombusviridae*) [96]. Additionally, the other smart technologies and high-throughput techniques have highly efficient agriculture analysis and can be integrated and applied with the innovation of artificial intelligence (AI) such as thermography [97], Raman spectroscopy (RS) [98], phytohormone biosensing and active remote sensing methods of radio detection and ranging (RADAR), and light detection and ranging (LiDAR) [99].

Plant EOs and their active compounds will be applied after preliminary detection and analysis by easy-to-use smart technologies in which the collected data was automated and real-time report. Therefore, the integration of different innovative technologies is providing for crop protection. Especially with smartphone applications that combine innovative technologies between imaging, telecommunications, and computing technologies including modern smartphones technologies, and smartphone-based volatile organic compound (VOC) sensor systems are interesting (**Figure 7**). Several free downloads of smartphone-based AI applications (crop diagnostics tools) can be applied with imaging and phenotyping for plant pathogen and insect pest identification such as Leaf Doctor [100], Pestoz, Plantix, PlantVillage Nuru, Agrio, PlantSnap, CropsAI, Plants Disease Identification, DoctorP, Crop Doctor, Purdue Tree Doctor, Leaf Plant Tech, and Tumaini. Li et al. [101] reported that the developable smartphone-based VOC fingerprinting platform with nanosensors and conventional chromogenic dyes was successfully detected the leaf volatile emissions at the early infection stage of *Phytophthora infestans* on tomato plants with the high detection accuracy of >95%. Similarly, several plant viruses were correctly detected by automated mobile apps such as *Banana bunchy top virus* (BBTV: *Babuvirus*, *Nanoviridae*) [102] and *Cassava brown streak virus* (CBSV: *Ipomovirus*, *Potyviridae*) [103].

7. Plant EO future challenges and perspectives under agriculture 4.0

The quality and stability of natural products depend on the quality of raw materials, extraction method techniques, and conditional storage. Therefore, the efficacy and role of natural extracts for antiviral and insect-pest management need to be considered as valuable and renewable processes. The research development of Agriculture 4.0 will improve the utilization of bioactive compounds. Crop protection under modern biotechnology collaborates innovative technologies in artificial intelligence to be used in the process design of extraction equipment and data storage. However, the success of Agriculture 4.0 will require policy and research support. Raising awareness of the value of natural products, the conservation of biodiversity and human health, and environmental safety will lead to the acceptance of agricultural products. This will create sustainability in modern farming systems.

The big data applications in smart farming can help the farmers in agricultural planning and executing activities to crop yield management. For example, integrated innovative technologies with software have the potential in detecting and monitoring plant diseases and insect pests. The smart network applications combined with the various push factors include general technological developments (e.g., IoT, AI, and agri-tech companies), sophisticated technologies (e.g., global navigation satellite systems, remote sensing, robots, and UAVs), data generation and storage, digital connectivity, and innovation possibilities that will enable efficiency for planning and operating agricultural works related to the pull factors (e.g., business and public drivers) [104]. This knowledge can be applied in the stages of the data and supply chains of plant EOs as follows: data capture, data storage, data transfer, data transformation, data analytics, and data marketing. Moreover, smart technical challenges and environmental treads related to security and safety as well as sustainability will be created, solved, and developed for big data in smart farming. The important issues for natural product development by innovative start-up companies are lacking many references for efficiency improvement, reliable quantitative analysis, and farmer's acceptance. The easily accessible platforms in real-time information are important for the benchmarking and modeling of business in supply chain scenarios and social media platforms. Integrated different players and partners in the short supply chains between the farmers and suppliers have potential management rather than integrated long supply chains. These operation models will reduce factors of the privacy and security of data ownership by the intelligent processing of management information systems. All of these are related to sustainable integration and smart-business models especially empower farmers and collaboration in all processes of supply chains through the openness of smart platforms. Consequently, the Plant EO future challenges under Agriculture 4.0 policy will be developed by the knowledge-based and knowledge engineering systems of integrated innovative technologies. The ultimate goals of developable natural product-based plant EOs and their active components with the various mechanisms of action will be designed for the farmers. As a result of these, the vital challenges will improve sustainable agricultural policies and strategies in the different crop systems under Agriculture 4.0.

8. Conclusion

The outbreak and resistance problems of plant pathogenic viruses and insect vectors to natural and chemical products have tended to increase. Farmers need safe and high-quality products to solve their problems. Several recent innovative technologies to develop and improve environmentally friendly products for antiviral and insect-pest management can be used to effectively control production quality under large-scale agricultural fields. Agriculture 4.0 is a modern model that can improve the efficiency of natural substances with the research development of extraction techniques and bioactive quality testing to promote and build farmers' acceptances. However, the modern agricultural system must be supported and cooperated by not only the government but also the public sectors to push the policy toward sustainable concrete practice for the highest benefit to environmentally friendly and humanity.

Acknowledgements

This book chapter was supported by Ph.D. research grants from the collaboration programs of Research and Researchers for Industries (RRi) of Thailand Science Research and Innovation (TSRI), Chia Tai Co., Ltd. (Contract No: PHD59I0061, Code: 5911004), and the Thammasat University Research Unit in Medicinal Chemistry, Thammasat University.

Conflict of interest

The authors declare no conflict of interest.

Author details

Thanat Na Phatthalung[1] and Wipa Tangkananond[2]*

1 Faculty of Science and Technology, Department of Biotechnology, Thammasat University, Pathum Thani, Thailand

2 Thammasat University Research Unit in Medicinal Chemistry, Thammasat University, Pathum Thani, Thailand

*Address all correspondence to: w_tangkananond@yahoo.com

References

[1] Zheng X. The uncertainties of population research: Challenges and opportunities. China CDC Weekly China CDC Wkly. 2021;**3**:591-592. DOI: 10.46234/ccdcw2021.156

[2] Zhang Q-W, Lin L-G, Ye W-C. Techniques for extraction and isolation of natural products: A comprehensive review. Chinese Medicine. 2018;**13**:20. DOI: 10.1186/s13020-018-0177-x

[3] Rivero A, Vézilier J, Weill M, Read AF, Gandon S. Insecticide control of vector-borne diseases: When is insecticide resistance a problem? PLoS Pathogens. 2010;**6**:e1001000. DOI: 10.1371/journal.ppat.1001000

[4] Liu Y, Pan X, Li J. A 1961-2010 record of fertilizer use, pesticide application and cereal yields: A review. Agronomy for Sustainable Development. 2015;**35**:83-93. DOI: 10.1007/s13593-014-0259-9

[5] Miresmailli S, Isman MB. Botanical insecticides inspired by plant-herbivore chemical interactions. Trends in Plant Science. 2014;**19**:29-35. DOI: 10.1016/j.tplants.2013.10.002

[6] Sulewski P, Kłoczko-Gajewska A, Sroka W. Relations between Agri-environmental, economic and social dimensions of farms' sustainability. Sustainability. 2018;**10**:4629. DOI: 10.3390/su10124629

[7] Redfern J, Kinninmonth M, Burdass D, Verran J. Using soxhlet ethanol extraction to produce and test plant material (essential oils) for their antimicrobial properties. Journal of Microbiology and Biology Education. 2014;**15**:45-46. DOI: 10.1128/jmbe.v15i1.656

[8] Kowalski R, Gagoś M, Kowalska G, Pankiewicz U, Sujka M, Mazurek A, et al. Effects of ultrasound technique on the composition of different essential oils. Journal of Analytical Methods in Chemistry. 2019;**6**:6782495. DOI: 10.1155/2019/6782495

[9] Božović M, Navarra A, Garzoli S, Pepi F, Ragno R. Essential oils extraction: A 24-hour steam distillation systematic methodology. Natural Product Research. 2017;**31**:2387-2396. DOI: 10.1080/14786419.2017.1309534

[10] Canbay HS. Effectiveness of liquid-liquid extraction, solid phase extraction, and headspace technique for determination of some volatile water-soluble compounds of rose aromatic water. International Journal of Analytical Chemistry. 2017;**2017**:4870671. DOI: 10.1155/2017/4870671

[11] Kamali H, Jalilvand MR, Aminimoghadamfarouj N. Pressurized fluid extraction of essential oil from *Lavandula hybrida* using a modified supercritical fluid extractor and a central composite design for optimization. Journal of Separation Science. 2012;**35**:1479-1485. DOI: 10.1002/jssc.201200043

[12] Pourmortazavi SM, Hajimirsadeghi SS. Supercritical fluid extraction in plant essential and volatile oil analysis. Journal of Chromatography. A. 2007;**1163**:2-24. DOI: 10.1016/j.chroma.2007.06.021

[13] Turrini F, Beruto M, Mela L, Curir P, Triglia G, Boggia R, et al. Ultrasound-assisted extraction of lavender (*Lavandula angustifolia* miller, cultivar Rosa) solid by-products remaining after the distillation of the essential oil. Applied Sciences. 2021;**11**:5495. DOI: 10.3390/app11125495

[14] Cardoso-Ugarte GA, Juárez-Becerra GP, Sosa-Morales ME,

López-Malo A. Microwave-assisted extraction of essential oils from herbs. The Journal of Microwave Power and Electromagnetic Energy. 2013;**47**:63-72. DOI: 10.1080/08327823.2013.11689846

[15] Ranjha MMAN, Kanwal R, Shafique B, Arshad RN, Irfan S, Kieliszek M, et al. A critical review on pulsed electric field: A novel technology for the extraction of phytoconstituents. Molecules. 2021;**26**:4893. DOI: 10.3390/molecules26164893

[16] Amudan R, Kamat D, Kamat S. Enzyme-assisted extraction of essential oils from *Syzygium aromaticum*. South Asian Journal of Experimental Biology. 2011;**1**:248-254

[17] Maes C, Bouquillon S, Fauconnier M-L. Encapsulation of essential oils for the development of biosourced pesticides with controlled release: A review. Molecules (Basel, Switzerland). 2019;**24**:2539. DOI: 10.3390/molecules24142539

[18] Lammari N, Louaer O, Meniai AH, Elaissari A. Encapsulation of essential oils via nanoprecipitation process: Overview, progress, challenges and prospects. Pharmaceutics. 2020;**12**:431. DOI: 10.3390/pharmaceutics12050431

[19] Cellini A, Blasioli S, Biondi E, Bertaccini A, Braschi I, Spinelli F. Potential applications and limitations of electronic nose devices for plant disease diagnosis. Sensors (Basel, Switzerland). 2017;**17**:2596. DOI: 10.3390/s17112596

[20] Riul A, de Sousa HC, Malmegrim RR, dos Santos DS, Carvalho ACPLF, Fonseca FJ, et al. Wine classification by taste sensors made from ultra-thin films and using neural networks. Sensors and Actuators, B: Chemical. 2004;**98**:77-82. DOI: 10.1016/j.snb.2003.09.025

[21] Wilson AD, Baietto M. Applications and advances in electronic-nose technologies. Sensors (Basel, Switzerland). 2009;**9**:5099-5148. DOI: 10.3390/s90705099

[22] Singh H, Raj VB, Kumar J, Mittal U, Mishra M, Nimal AT, et al. Metal oxide SAW E-nose employing PCA and ANN for the identification of binary mixture of DMMP and methanol. Sensors and Actuators, B: Chemical. 2014;**200**: 147-156. DOI: 10.1016/j.snb.2014.04.065

[23] Daneshkhah A, Vij S, Siegel AP, Agarwal M. Polyetherimide/carbon black composite sensors demonstrate selective detection of medium-chain aldehydes including nonanal. Chemical Engineering Journal. 2020;**383**:123104. DOI: 10.1016/j.cej.2019.123104

[24] Karakaya D, Ulucan O, Turkan M. Electronic nose and its applications: A survey. International Journal of Automation and Computing. 2020;**17**:179-209. DOI: 10.1007/s11633-019-1212-9

[25] Wilson AD. Review of electronic-nose technologies and algorithms to detect hazardous chemicals in the environment. Procedia Technology. 2012;**1**:453-463. DOI: 10.1016/j.protcy.2012.02.101

[26] Shao X, Li H, Wang N, Zhang Q. Comparison of different classification methods for analyzing electronic nose data to characterize sesame oils and blends. Sensors (Basel, Switzerland). 2015;**15**:26726-26742. DOI: 10.3390/s151026726

[27] Tan J, Xu J. Applications of electronic nose (e-nose) and electronic tongue (e-tongue) in food quality-related properties determination: A review. Artificial Intelligence in Agriculture. 2020;**4**:104-115. DOI: 10.1016/j.aiia.2020.06.003

[28] Rasekh M, Karami H, Wilson AD, Gancarz M. Performance analysis

of MAU-9 electronic-nose MOS sensor array components and ANN classification methods for discrimination of herb and fruit essential oils. Chemosensors. 2021;**9**:243. DOI: 10.3390/chemosensors9090243

[29] Rasekh M, Karami H, Wilson AD, Gancarz M. Classification and identification of essential oils from herbs and fruits based on a MOS electronic-nose technology. Chemosensors. 2021;**9**:142. DOI: 10.3390/chemosensors9060142

[30] Okur S, Sarheed M, Huber R, Zhang Z, Heinke L, Kanbar A, et al. Identification of mint scents using a QCM based e-nose. Chemosensors. 2021;**9**:31. DOI: 10.3390/chemosensors9020031

[31] Gómez AH, Wang J, Hu G, Pereira AG. Electronic nose technique potential monitoring mandarin maturity. Sensors and Actuators, B: Chemical. 2006;**113**:347-353. DOI: 10.1016/j.snb.2005.03.090

[32] Hong X, Wang J, Qi G. E-nose combined with chemometrics to trace tomato-juice quality. Journal of Food Engineering. 2015;**149**:38-43. DOI: 10.1016/j.jfoodeng.2014.10.003

[33] Lashgari M, MohammadiGol R. Discrimination of Golab apple storage time using acoustic impulse response and LDA and QDA discriminant analysis techniques. Iran Agricultural Research. 2016;**35**:65-70

[34] Graboski AM, Zakrzevski CA, Shimizu FM, Paschoalin RT, Soares AC, Steffens J, et al. Electronic nose based on carbon nanocomposite sensors for clove essential oil detection. ACS Sensors. 2020;**5**:1814-1821. DOI: 10.1021/acssensors.0c00636

[35] Lias S, Mohamad Ali NA, Jamil M, Tolmanan MSY, Misman MA. A study on the application of electronic nose coupled with DFA and statistical analysis for evaluating the relationship between sample volumes versus sensor intensity of agarwood essential oils blending ratio. MATEC Web of Conferences. 2018;**201**:02008. DOI: 10.1051/matecconf/201820102008

[36] Wu S-Q, Li R, Jiang Z-T, Wang Y, Tan J, Tang S-H. Evaluation of antioxidant active ingredients of spikenard essential oil by ultra-fast gas chromatography electronic nose and radical scavenging mechanism. Industrial Crops and Products. 2020;**151**:112489. DOI: 10.1016/j.indcrop.2020.112489

[37] Abbey L, Aked J, Joyce DC. Discrimination amongst *alliums* using an electronic nose. The Annals of Applied Biology. 2001;**139**:337-342. DOI: 10.1111/J.1744-7348.2001.TB00147.X

[38] Mamatha BS, Prakash M. Studies on pepper (*Piper nigrum* L.) cultivars by sensory and instrumental techniques. Z Arznei- Gewurzpflanzen. 2011;**16**:176-180

[39] Zawirska-Wojtasiak R, Gośliński M, Szwacka M, Gajc-Wolska J, Mildner-Szkudlarz S. Aroma evaluation of transgenic, thaumatin II-producing cucumber fruits. Journal of Food Science. 2009;**74**:C204-C210. DOI: 10.1111/j.1750-3841.2009.01082.x

[40] Sarowska J, Wojnicz D, Jama-Kmiecik A, Frej-Mądrzak M, Choroszy-Król I. Antiviral potential of plants against noroviruses. Molecules (Basel, Switzerland). 2021;**26**:4669. DOI: 10.3390/molecules26154669

[41] Bishop CD. Antiviral activity of the essential oil of *Melaleuca alternifolia* (maiden and Betche) Cheel (tea tree) against *tobacco mosaic virus*. Journal of Essential Oil Research. 1995;7:641-644. DOI: 10.1080/10412905.1995.9700519

[42] Iftikhar S, Shahid AA, Javed S, Nasir IA, Tabassum B, Haider M. Essential oils and latices as novel antiviral agent against *potato leaf roll virus* and analysis of their phytochemical constituents responsible for antiviral activity. The Journal of Agricultural Science. 2013;**5**:167-188. DOI: 10.5539/jas.v5n7p167

[43] Lu M, Han Z, Xu Y, Yao L. *In vitro* and *in vivo* anti-*tobacco mosaic virus* activities of essential oils and individual compounds. Journal of Microbiology and Biotechnology. 2013;**23**:771-778. DOI: 10.4014/jmb.1210.10078

[44] Dikova B, Dobreva A, Djurmanski A. Essential oils of lavender and fennel for inhibiting *tomato spotted wilt virus* in pepper plants. Acta Microbiol Bulg. 2017;**33**:36-43

[45] Rao GP, Pandey AK, Shukla K. Essential oils of some higher plants Vis-a-Vis some legume viruses. Indian Perfumer. 1986;**30**:483-486

[46] Helal IM. Use of biocides for controlling viral diseases that attack common bean and cucumber plants. Folia Horticulturae. 2019;**31**:159-170. DOI: 10.2478/fhort-2019-0011

[47] Na Phatthalung T, Tangkananond W. Brown planthopper vector-virus transmission in rice and inhibitory effects of plant essential oils. International Journal of Agricultural Technology. 2021;**17**:587-606

[48] Cagno V, Donalisio M, Civra A, Cagliero C, Rubiolo P, Lembo D. *In vitro* evaluation of the antiviral properties of Shilajit and investigation of its mechanisms of action. Journal of Ethnopharmacology. 2015;**166**:129-134. DOI: 10.1016/j.jep.2015.03.019

[49] Aoki-Utsubo C, Chen M, Hotta H. Time-of-addition and temperature-shift assays to determine particular step(s) in the viral life cycle that is blocked by antiviral substance(s). Bio-Protocol. 2018;**8**:e2830. DOI: 10.21769/BioProtoc.2830

[50] Panikar S, Shoba G, Arun M, Sahayarayan JJ, Usha Raja Nanthini A, Chinnathambi A, et al. Essential oils as an effective alternative for the treatment of COVID-19: Molecular interaction analysis of protease (M^{pro}) with pharmacokinetics and toxicological properties. Journal of Infection and Public Health. 2021;**14**:601-610. DOI: 10.1016/j.jiph.2020.12.037

[51] Ma L, Yao L. Antiviral effects of plant-derived essential oils and their components: An updated review. Molecules (Basel, Switzerland). 2020;**25**:2627. DOI: 10.3390/molecules25112627

[52] Abou Baker DH, Amarowicz R, Kandeil A, Ali MA, Ibrahim EA. Antiviral activity of *Lavandula angustifolia* L. and *Salvia officinalis* L. essential oils against avian influenza H5N1 virus. Journal of Agriculture and Food Research. 2021;**4**:100135. DOI: 10.1016/j.jafr.2021.100135

[53] Siejak P, Smułek W, Fathordobady F, Grygier A, Baranowska HM, Rudzińska M, et al. Multidisciplinary studies of folk medicine “five thieves’ oil” (Olejek Pięciu Złodziei) components. Molecules. 2021;**26**:2931. DOI: 10.3390/molecules26102931

[54] Singh N, Wang C, Cooper R. Potential of essential oil-based pesticides and detergents for bed bug control. Journal of Economic Entomology. 2014;**107**:2163-2170. DOI: 10.1603/ec14328

[55] Na Phatthalung T, Tangkananond W. *Rice grassy stunt virus*-free and pathogenic rice plants affect the brown planthopper (*Nilaparvata lugens* Stål) life cycle.

Agriculture and Natural Resources. 2021;**55**:331-340. DOI: 10.34044/j.anres.2021.55.3.02

[56] Shafie R, Kheder A, Farghaly A. Induction of resistance in pepper plants against *potato virus Y* (PVY)^{NTN} by two medicinal and aromatic plant essential oils and their major components. Egyptian Journal of Phytopathology. 2017;**45**:1-15. DOI: 10.21608/ejp.2017.89458

[57] Saiz-Rubio V, Rovira-Más F. From smart farming towards agriculture 5.0: A review on crop data management. Agronomy. 2020;**10**:207. DOI: 10.3390/agronomy10020207

[58] Janssen SJC, Porter CH, Moore AD, Athanasiadis IN, Foster I, Jones JW, et al. Towards a new generation of agricultural system data, models and knowledge products: Information and communication technology. Agricultural Systems. 2017;**155**:200-212. DOI: 10.1016/j.agsy.2016.09.017

[59] Shivappa H, Prakasa Rao EVS, Gouda KC, Ramesh K, Rakesh V, Mohapatra GN, et al. Digital revolution and big data: A new revolution in agriculture. CAB Reviews. 2018;**13**:021. DOI: 10.1079/PAVSNNR201813021

[60] Sørensen CG, Fountas S, Nash E, Pesonen L, Bochtis D, Pedersen SM, et al. Conceptual model of a future farm management information system. Computers and Electronics in Agriculture. 2010;**72**:37-47. DOI: 10.1016/j.compag.2010.02.003

[61] Salami P, Ahmadi H. Review of farm management information systems (FMIS). New York Science Journal. 2010;**3**:87-95

[62] Verdouw CN, Robbemond RM, Wolfert J. ERP in agriculture: Lessons learned from the Dutch horticulture. Computers and Electronics in Agriculture. 2015;**114**:125-133. DOI: 10.1016/j.compag.2015.04.002

[63] Silvie PJ, Martin P, Huchard M, Keip P, Gutierrez A, Sarter S. Prototyping a knowledge-based system to identify botanical extracts for plant health in sub-Saharan Africa. Plants. 2021;**10**:896. DOI: 10.3390/plants10050896

[64] Pantazi XE, Moshou D, Oberti R, West J, Mouazen AM, Bochtis D. Detection of biotic and abiotic stresses in crops by using hierarchical self organizing classifiers. Precision Agriculture. 2017;**18**:383-393. DOI: 10.1007/s11119-017-9507-8

[65] Hashimoto N, Saito Y, Maki M, Homma K. Simulation of reflectance and vegetation indices for unmanned aerial vehicle (UAV) monitoring of paddy fields. Remote Sensing. 2019;**11**:2119. DOI: 10.3390/rs11182119

[66] Murugan D, Garg A, Singh D. Development of an adaptive approach for precision agriculture monitoring with drone and satellite data. IEEE Journal of Selected Topics in Applied Earth Observations and Remote Sensing. 2017;**10**:5322-5328. DOI: 10.1109/JSTARS.2017.2746185

[67] García-Martínez H, Flores-Magdaleno H, Khalil-Gardezi A, Ascencio-HernándezR,Tijerina-ChávezL, Vázquez-Peña MA, et al. Digital count of corn plants using images taken by unmanned aerial vehicles and cross correlation of templates. Agronomy. 2020;**10**:469. DOI: 10.3390/agronomy10040469

[68] Mohidem NA, Che'Ya NN, Juraimi AS, Fazlil Ilahi WF, Mohd Roslim MH, Sulaiman N, et al. How can unmanned aerial vehicles be used for detecting weeds in agricultural fields? Agriculture 2021;**11**:1004. DOI: 10.3390/agriculture11101004

[69] Kachamba DJ, Ørka HO, Gobakken T, Eid T, Mwase W. Biomass estimation using 3D data from unmanned aerial vehicle imagery in a tropical woodland. Remote Sensing. 2016;**8**:968. DOI: 10.3390/rs8110968

[70] Zhang W, Li X, Yu J, Kumar M, Mao Y. Remote sensing image mosaic technology based on SURF algorithm in agriculture. EURASIP Journal on Image and Video Processing. 2018;**2018**:85. DOI: 10.1186/s13640-018-0323-5

[71] Wiesner-Hanks T, Wu H, Stewart E, DeChant C, Kaczmar N, Lipson H, et al. Millimeter-level plant disease detection from aerial photographs *via* deep learning and crowdsourced data. Frontiers in Plant Science. 2019;**10**:1550. DOI: 10.3389/fpls.2019.01550

[72] Wang G, Lan Y, Qi H, Chen P, Hewitt A, Han Y. Field evaluation of an unmanned aerial vehicle (UAV) sprayer: Effect of spray volume on deposition and the control of pests and disease in wheat. Pest Management Science. 2019;**75**:1546-1555. DOI: 10.1002/ps.5321

[73] Tillett ND, Hague T. Increasing work rate in vision guided precision banded operations. Biosystems Engineering. 2006;**94**:487-494. DOI: 10.1016/j.biosystemseng.2006.04.010

[74] Mongkolchart N, Ketcham M. The surveillance system for rice diseases detection using color model. International Journal of Computer Integrated Manufacturing. 2020;**28**:26-35

[75] Xie C, Shao Y, Li X, He Y. Detection of early blight and late blight diseases on tomato leaves using hyperspectral imaging. Scientific Reports. 2015;**5**:16564. DOI: 10.1038/srep16564

[76] Zhu H, Chu B, Zhang C, Liu F, Jiang L, He Y. Hyperspectral imaging for presymptomatic detection of tobacco disease with successive projections algorithm and machine-learning classifiers. Scientific Reports. 2017;**7**:4125. DOI: 10.1038/s41598-017-04501-2

[77] Jin X, Jie L, Wang S, Qi HJ, Li SW. Classifying wheat hyperspectral pixels of healthy heads and fusarium head blight disease using a deep neural network in the wild field. Remote Sensing. 2018;**10**:395. DOI: 10.3390/rs10030395

[78] Kanzi AM, San JE, Chimukangara B, Wilkinson E, Fish M, Ramsuran V, et al. Next generation sequencing and bioinformatics analysis of family genetic inheritance. Frontiers in Genetics. 2020;**11**:544162. DOI: 10.3389/fgene.2020.544162

[79] Notomi T, Okayama H, Masubuchi H, Yonekawa T, Watanabe K, Amino N, et al. Loop-mediated isothermal amplification of DNA. Nucleic Acids Research. 2000;**28**:E63-E63. DOI: 10.1093/nar/28.12.e63

[80] Ambrico M, Ambrico PF, Minafra A, De Stradis A, Vona D, Cicco SR, et al. Highly sensitive and practical detection of plant viruses via electrical impedance of droplets on textured silicon-based devices. Sensors. 2016;**16**:1946. DOI: 10.3390/s16111946

[81] Boukhatem MN, Setzer WN. Aromatic herbs, medicinal plant-derived essential oils, and phytochemical extracts as potential therapies for coronaviruses: Future perspectives. Plants (Basel). 2020;**9**:800. DOI: 10.3390/plants9060800

[82] Dunkić V, Bezić N, Vuko E, Cukrov D. Antiphytoviral activity of *Satureja montana* L. ssp. variegata (host) P. W. Ball essential oil and phenol compounds on CMV and TMV. Molecules. 2010;**15**:6713-6721. DOI: 10.3390/molecules15106713

[83] Sun W-J, Lv W-J, Li L-N, Yin G, Hang X, Xue Y, et al. Eugenol confers resistance to *tomato yellow leaf curl virus* (TYLCV) by regulating the expression of SlPer1 in tomato plants. New Biotechnology. 2016;**33**:345-354. DOI: 10.1016/j.nbt.2016.01.001

[84] Sharifi-Rad J, Salehi B, Schnitzler P, Ayatollahi SA, Kobarfard F, Fathi M, et al. Susceptibility of *herpes simplex virus* type 1 to monoterpenes thymol, carvacrol, p-cymene and essential oils of *Sinapis arvensis* L., *Lallemantia royleana* Benth. And *Pulicaria vulgaris* Gaertn. Cellular and Molecular Biology. 2017;**63**:42-47. DOI: 10.14715/cmb/2017.63.8.10

[85] Silva-Flores PG, Pérez-López LA, Rivas-Galindo VM, Paniagua-Vega D, Galindo-Rodríguez SA, Álvarez-Román R. Simultaneous GC-FID quantification of main components of *Rosmarinus officinalis* L. and *Lavandula dentata* essential oils in polymeric nanocapsules for antioxidant application. Journal of Analytical Methods in Chemistry. 2019;**2019**:2837406. DOI: 10.1155/2019/2837406

[86] Tholl D, Hossain O, Weinhold A, Röse USR, Wei Q. Trends and applications in plant volatile sampling and analysis. The Plant Journal. 2021;**106**:314-325. DOI: 10.1111/tpj.15176

[87] Wang D, Vinson R, Holmes M, Seibel G, Bechar A, Nof S, et al. Early detection of *tomato spotted wilt virus* by hyperspectral imaging and outlier removal auxiliary classifier generative adversarial nets (OR-AC-GAN). Scientific Reports. 2019;**9**:4377. DOI: 10.1038/s41598-019-40066-y

[88] Zhu H, Cen H, Zhang C, He Y. Early detection and classification of tobacco leaves inoculated with *tobacco mosaic virus* based on hyperspectral imaging technique. ASABE Annual International Meeting. 2016;**3**:1862-1868. DOI: 10.13031/aim.20162460422

[89] Nguyen C, Sagan V, Maimaitiyiming M, Maimaitijiang M, Bhadra S, Kwasniewski MT. Early detection of plant viral disease using hyperspectral imaging and deep learning. Sensors (Basel). 2021;**21**:742. DOI: 10.3390/s21030742

[90] Polder G, van der Heijden GWAM, van Doorn J, Baltissen TAHMC. Automatic detection of *tulip breaking virus* (TBV) in tulip fields using machine vision. Biosystems Engineering. 2014;**117**:35-42. DOI: 10.1016/j.biosystemseng.2013.05.010

[91] Polder G, Blok PM, de Villiers HAC, van der Wolf JM, Kamp J. *Potato virus Y* detection in seed potatoes using deep learning on hyperspectral images. Frontiers in Plant Science. 2019;**10**:209. DOI: 10.3389/fpls.2019.00209

[92] Han XY, Li PX, Zou LJ, Tan WR, Zheng T, Zhang DW, et al. GOLDEN2-LIKE transcription factors coordinate the tolerance to *cucumber mosaic virus* in Arabidopsis. Biochemical and Biophysical Research Communications. 2016;**477**:626-632. DOI: 10.1016/j.bbrc.2016.06.110

[93] Chaerle L, Leinonen I, Jones HG, Van Der Straeten D. Monitoring and screening plant populations with combined thermal and chlorophyll fluorescence imaging. Journal of Experimental Botany. 2007;**58**:773-784. DOI: 10.1093/jxb/erl257

[94] Chaerle L, Pineda M, Romero-Aranda R, Van Der Straeten D, Barón M. Robotized thermal and chlorophyll fluorescence imaging of *pepper mild mottle virus* infection in *Nicotiana benthamiana*. Plant & Cell Physiology. 2006;**47**:1323-1336. DOI: 10.1093/pcp/pcj102

[95] Wang L, Poque S, Valkonen JPT. Phenotyping viral infection in sweetpotato using a high-throughput chlorophyll fluorescence and thermal imaging platform. Plant Methods. 2019;**15**:116. DOI: 10.1186/s13007-019-0501-1

[96] Pu XJ, Li YN, Wei LJ, Xi DH, Lin HH. Mitochondrial energy-dissipation pathway and cellular redox disruption compromises Arabidopsis resistance to *turnip crinkle virus* infection. Biochemical and Biophysical Research Communications. 2016;**473**:421-427. DOI: 10.1016/j.bbrc.2016.03.023

[97] Pineda M, Barón M, Pérez-Bueno M-L. Thermal imaging for plant stress detection and phenotyping. Remote Sensing. 2021;**13**:68. DOI: 10.3390/rs13010068

[98] Farber C, Bryan R, Paetzold L, Rush C, Kurouski D. Non-invasive characterization of single-, double- and triple-viral diseases of wheat with a hand-held Raman spectrometer. Frontiers in Plant Science. 2020;**11**:01300. DOI: 10.3389/fpls.2020.01300

[99] Cubero S, Marco-Noales E, Aleixos N, Barbé S, Blasco J. RobHortic: A field robot to detect pests and diseases in horticultural crops by proximal sensing. Agriculture. 2020;**10**:276. DOI: 10.3390/agriculture10070276

[100] Pethybridge SJ, Nelson SC. Leaf doctor: A new portable application for quantifying plant disease severity. Plant Disease. 2015;**99**:1310-1316. DOI: 10.1094/pdis-03-15-0319-re

[101] Li Z, Paul R, Ba Tis T, Saville AC, Hansel JC, Yu T, et al. Non-invasive plant disease diagnostics enabled by smartphone-based fingerprinting of leaf volatiles. Nature Plants. 2019;**5**:856-866. DOI: 10.1038/s41477-019-0476-y

[102] Selvaraj MG, Vergara A, Ruiz H, Safari N, Elayabalan S, Ocimati W, et al. AI-powered banana diseases and pest detection. Plant Methods. 2019;**15**:92. DOI: 10.1186/s13007-019-0475-z

[103] Mrisho LM, Mbilinyi NA, Ndalahwa M, Ramcharan AM, Kehs AK, McCloskey PC, et al. Accuracy of a smartphone-based object detection model, PlantVillage Nuru, in identifying the foliar symptoms of the viral diseases of cassava-CMD and CBSD. Frontiers in Plant Science. 2020;**11**:590889. DOI: 10.3389/fpls.2020.590889

[104] Wolfert S, Ge L, Verdouw C, Bogaardt M-J. Big data in smart farming – A review. Agricultural Systems. 2017;**153**: 69-80. DOI: 10.1016/j.agsy.2017.01.023

Section 2

Application

Chapter 3

Autonomous Aerial Robotic System for Smart Spraying Tasks: Potentials and Limitations

Petar Piljek, Marko Pranjić, Denis Kotarski and Tomislav Petanjek

Abstract

Continuous demands for growth in agricultural productivity and modern demands for the sustainable agricultural approach are bringing farmers into a new technological era. With all the limitations and risks, precision agriculture and other related technologies show great potential in solving the challenges of sustainable and more efficient agricultural production. Nowadays, unmanned aerial vehicles (UAVs) are able to perform a wide range of agricultural tasks, from data collection to smart spraying. This chapter presents the concept of a modular autonomous robotic system that, based on available technologies, materials, and system components, can be produced and applied in precision agriculture. The primary purpose of such a system, which consists of a multirotor UAV and docking station, is to save the time required to perform the task and to reduce environmental and soil pollution. Several problems have been addressed, which affect performance and energy consumption, for example, of spraying a field crop.

Keywords: precision agriculture, multirotor UAV, modular system, smart spraying task

1. Introduction

Agricultural production has continuously progressed from primitive techniques and tools to modern comprehensive digitized processes and systems. This evolutionary process can be presented in four main steps from Agriculture 1. To Agriculture 4.0. Agriculture 1.0 is based on simple tools, manpower, and animal forces and can be placed up to the nineteenth century. Agriculture 2.0 follows first industrial revolution and introduces various agricultural machinery operated by farmers and use of plenty chemicals. Agriculture 3.0 emerged in the twentieth century through the usage of automation and robotic techniques thanks to the rise of information and communication technologies (ICTs). Production became more efficient, and some environmental problems were reduced. In the present day, the main aims of Agriculture 4.0 are associated with the introduction of further automation and new digital technologies such as Internet of things (IoT), big data, artificial intelligence (AI), remote sensing, cloud computing, wireless sensor network in agriculture production, allowing a transition toward smart and sustainable farming. This advanced automation and process digitalization have resulted in emergence of the precision agriculture (PA), a farming management concept that utilizes the available technology with aims to improve productivity, efficiency and profitability, quality of the crops and product, along with sustainability and the protection of the environment. Although the principles of PA have been known for more than 25 years, they became interesting to farmers in the last decade due to technological advances and the adoption of new technologies. Thanks to intensive research and technological advances, unmanned aerial vehicles (UAVs) have also undergone through tremendous technical progress over the last decade, which is why they are used today to perform a variety of tasks in many industries. The global agriculture unmanned aerial vehicles (UAVs) market is expected to reach 5,7 billion of USD by 2025. One of the promising areas of application is also the use of UAVs in PA where they are used for a whole range of tasks, from data collection to smart spraying tasks. The utilization of various technologies in PA has been extensively researched and documented in several scientific papers. Nowadays, some of the key terms related to PA are remote sensing, automated hardware, control systems, software, global positioning system (GPS) guidance, robotics, unmanned ground vehicles (UGVs), UAV, and so on.

Information technologies (ITs) used in PA and criteria for their comparison and selection, to store, recover, transmit, and manipulate agricultural data are identified in [1]. The identified IT are GPS, multimedia devices (devices that allow capturing images or videos, such as smartphones or cameras), nano sensors, remote sensors, sensors in general, unmanned aerial systems (UASs), UAV, UGV, variable rate technology (VRT), and wireless sensor networks (WSNs). A survey given in [2] includes wireless communication technologies, sensors, and wireless nodes used to assess the environmental behavior, the platforms used to obtain spectral images of crops, the common vegetation indices used to analyze spectral images, and applications of WSN in agriculture. Authors have also proposed a smart solution for crop health monitoring based on the Internet of things (IoT) and comprising two modules, the wireless sensor network–based system to monitor real-time crop health status and a low-altitude remote sensing platform to obtain multispectral imagery. The work [3] deals with the influence of the fourth industrial revolution on PA. The revolution is expected to spur new technological innovation in six areas: artificial intelligence, robotics, IoT, unmanned vehicles, three-dimensional printing, and nanotechnology. Additionally, it will include a range of new technologies that use big data to incorporate the physical, biological, and digital worlds. Detailed analysis of UAV applications for PA is given [4], where all applications are divided into three categories: UAV-based monitoring applications, UAV-based spraying applications, and multi-UAV applications where multiple UAVs are used to accomplish a task. The application of small UAS for mapping and monitoring in PA is discussed in [5].

PA must quantify variations in soil and crop within agricultural fields, hence the following works also discuss various remote sensing technologies [6], sensor fusion [7], and deep learning techniques [8] to be able to automate processes and make decisions based on the sensor readings. Some research papers also deal with specific types of corps, such as orchard management [9], monitoring of nitrogen status of potatoes [10], detecting green weeds in preharvest cereals [11], and rice [12]. The main driver of PA was tractor GPS technology, which enabled accurate terrain mapping and meeting individual crop needs with different dosages of pesticides for different areas, depending on the reading from different sensors that can be fixed or mobile. Nowadays, ground vehicles of various types, sizes, and power sources are used to accomplish various tasks for PA purposes. Design and field evaluation of a ground robot as a new phenotyping platform that can measure individual plant architecture traits accurately over large areas at a subdaily frequency is demonstrated in [13]. Autonomous mobile robot based on a commercial agricultural vehicle chassis as a robotized patch sprayer is presented in [14], while in [15], the development of a small electrical robot intended to use for autonomous spraying is shown. In work [16] solar-powered UGV is presented that has multiple degrees of freedom positioning mechanism, and it is equipped with a robotic arm and vision sensors, which allow to challenge irregular terrains and to perform precision field operations with perception. There are many applications of solar systems used in agricultural production, and some are listed in the paper [17]. Numerous studies have been conducted, which consider heterogeneous robotic systems, mainly combinations of UGV and UAV. Ground and aerial measurements used for estimating nitrogen levels on-demand across a farm are presented in [18]. The main tasks of UGV in the context of UAV-UGV cooperation are considered in research [19]. The capability of images acquired from UAVs with multispectral cameras to detect weed patches and to support herbicide patch spraying is presented in [20]. Furthermore, the research [21] described a fleet of heterogeneous ground and aerial robots, developed, and equipped with innovative sensors, enhanced end effectors, and improved decision control algorithms to cover a large variety of agricultural situations.

UAVs have been used in a wide range of applications to support digital agriculture, including field scouting [22], precision management of oil palm plantation [23, 24], estimating plant's parameters such as leaf area index and height [25], health assessment [26], and variable rate spraying [27, 28]. The technologies of aerial electrostatic spraying using UAVs are being investigated [29], as well as the development of automatic aerial spraying systems based on UAVs [30, 31]. The design of an embedded real-time UAV spraying control system, based on low-cost hardware, which supports onboard image processing, is proposed in [32]. The use of computer-controlled swarms of UAVs for crop spraying enables nonuniform coverage of high precision and time efficiency, therefore an algorithmic control method for autonomous UAV swarm spraying is proposed in [33]. The static configuration usually adopted in the literature deals with the development of spraying processes have shortcomings in terms of changing weather conditions (e.g., sudden changes of wind speed and direction). To overcome this deficiency, in paper [34], an adaptive approach for UAV-based pesticide spraying in dynamic environments is presented. Also, in the paper [35], an algorithm for adjustment of the UAV route with respect to changes in wind intensity and direction is described, input of which is the feedback obtained from the WSN deployed in the crop field. Furthermore, the influence of windward airflow and droplet size on the movement of droplet groups is investigated. In [36], a numerical simulation and computational fluid dynamics analysis on spray drift movement are conducted for multirotor UAVs. Since the different spray requirements are possible, the variable spray system, which can rapidly adjust the flow range of the nozzle, is presented in [37]. The key problem in the task of smart spraying using drones is the distribution of droplets, so many scientific papers have been published on this topic [38–40].

In this chapter, a concept of an autonomous aerial robotic system intended for smart spraying tasks is presented. The presented system consists of a mobile base station and a multirotor UAV armed with spray equipment and a spraying tank. The main purpose of the concept is autonomous execution of spraying tasks on parcels of different surface ranges. The advantages and current problems related to the use of UAVs in smart spraying tasks are stated, and guidelines for the design of the base station are given.

Since multirotor UAVs are characterized by high energy consumption, special emphasis is placed on the characterization and adequate selection of components in order to obtain satisfactory flight performance and necessary flight duration. Furthermore, the aircraft system is divided into four subsystems (equipment and payload, electric energy, electric propulsion, and control subsystem), thus achieving a certain degree of modularity. In the last part of the paper, guidelines for designing a real system through the phases of characterization, analysis, and simulation are presented.

2. Precision agriculture: UAV integration

UAVs are found in a wide range of applications in PA due to their advantages over the use of current agricultural machinery. Their flexibility and a high degree of autonomy, along with low labor needs and avoidance of crops and soil damage, significantly increase agricultural productivity and sustainability. The efficient use of chemicals in agricultural production is crucial in order to reduce harm to human health and also to reduce costs. UAVs can be an effective and inexpensive alternative to conventional spraying, and applications can be extended to crop fertilization, seed sowing, and similar activities. The equipment in charge of spraying can be relatively easily retrofitted to this type of aircraft, which further reduces the cost of the system. In terms of system autonomy, a multirotor type of UAV is able to perform precision pesticide spraying missions given the specifics of the crop, the severity of the disease or pest, the location, and other requirements. The key thing in carrying out the mission is precisely controlled droplets deposition on the target and reducing the environmental pollution. Several UAV system parameters need to be considered, including flight route (path pattern), spraying height, flight speed, nozzle flow rate, number and orientation of nozzles, and others. There are several commercial smart spraying systems, and one of the most used all-in-one solutions is DJI Agras (**Figure 1**) [41].

Multirotor unmanned aerial vehicles intended for plant protection can be used on flat plots but also hilly and extremely uneven terrain. The application of an aerial robotic system for smart spraying missions in the rural area of Hrvatsko Zagorje, which is characterized by hilly terrain (relief), was considered, where typical landscape is shown in **Figure 2**. Apart from the demanding terrain, the problem is the fragmentation of plots and an uneven distribution of crops (by square footage and shape). Besides, some plots are very difficult to access with the machinery currently in use because there are very narrow roads between plots that are often unorganized, and some plots do not have any access roads. The abovementioned implies the need to design a flexible robotic system that can be used on parcels of wider square footage. In this chapter, the concept of an aerial robotic system consisting of a mobile base station and a multirotor UAV armed with spray equipment and a tank is considered. The possibility of performing vertical take-off and landing of a multirotor type of UAV allows easy docking of the aircraft with the base station.

A base station is a mobile multifunctional docking facility that has several functions. From the aspect of system planning and control, the essential component is a computer with associated modules that send and receive wireless signals from the aircraft online and also serve as an interface between the user and the aircraft. The mission parameters can be set via the base station, i.e., the flight can be planned based on the tasks that the aircraft needs to perform. The base station will determine flight parameters (path, speed, height) based on the required pesticide amount for specific area and the volume of spraying tank. Mission parameters determine the course of

Figure 1.
DJI Agras MG-1 commercial aircraft [42].

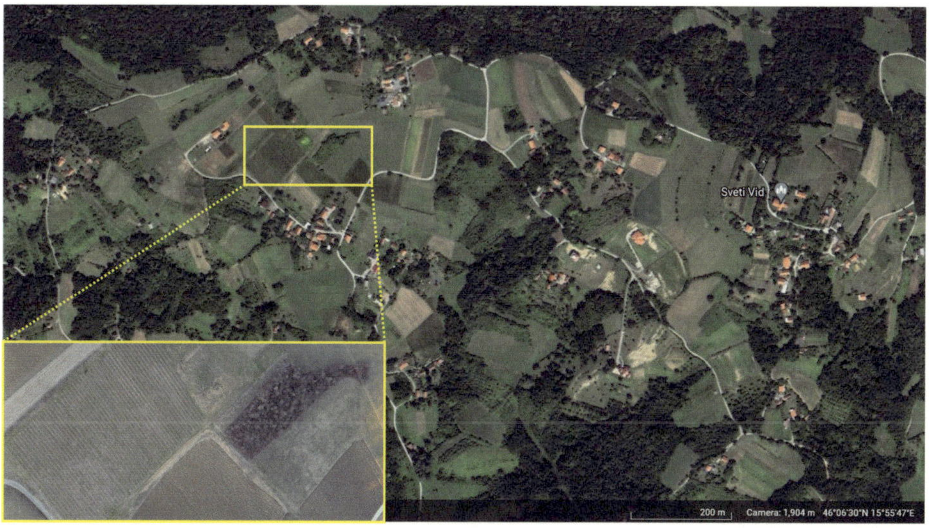

Figure 2.
Presentation of a typical landscape in Hrvatsko Zagorje characterized by small and irregular plots.

execution since this type of system can be used for different dimensions of plots and can also be used to perform a task on several plots. The base station should be able to change the batteries as needed for the mission and recharge the tank. After the aircraft completes the first part of the task and consumes the chemical, it returns vertically to the base station to fill up the tank and replace the battery. After the change, the aircraft performs a vertical take-off and continues to perform the task of spraying at the place where it stopped before loading. It follows from the above mentioned that the base station must be designed in such a way as to enable aircraft take-off

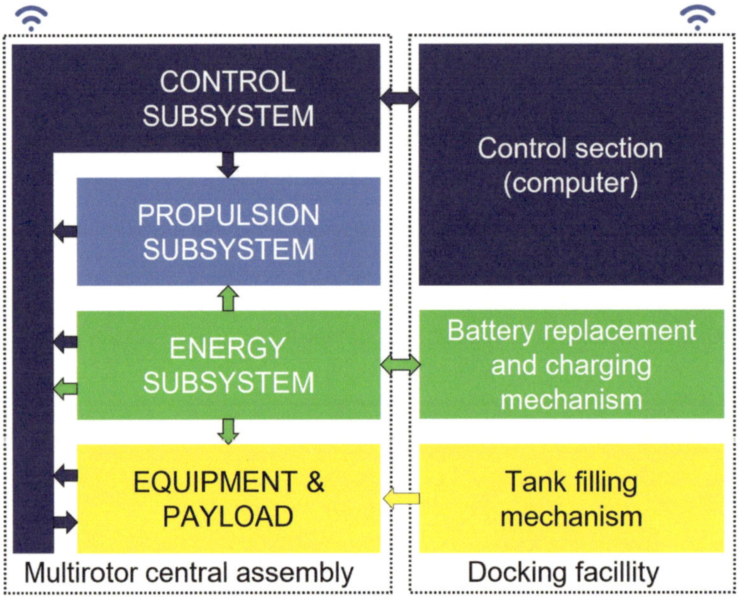

Figure 3.
Schematic representation of the concept of an aerial robotic system.

and landing, two-way communication, easy and safe replacement of batteries, and pump to fill up the tank. In addition to the listed basic functions, the base station can also have a module (generator) for charging batteries. **Figure 3** schematically shows the concept of an autonomous aerial robotic system consisting of a multifunctional mobile base station and a multirotor aircraft for smart spraying tasks.

3. Aerial robot system description

Multirotor aircraft are mechanical systems that exist in 3D space with six degrees of freedom (DOF) consisting of N rotors. From the aspect of dynamics, they are considered as symmetrical rigid bodies, where the only moving parts are the rotors of the propulsion assembly on whose axes are mounted propellers with a fixed pitch angle. Propellers create aerodynamic forces and moments by their rotation, so it follows that the angular velocities of the rotor are the only variables that have a direct impact on flight dynamics. The development and design of multirotor UAVs depend on constraints in size and energy consumption, and a key parameter in system design is aircraft weight. Given that the multirotor type of UAV is characterized by high energy consumption, it is extremely important to correctly select the components and parameters of the system in order to reduce energy consumption and extend the flight duration. To ensure overall flight performance, it is necessary to determine the thrust-to-weight ratio (TWR), and as a rule, aircraft are designed with approximately twice the thrust of the weight.

3.1 Equipment and payload subsystem

The equipment of a multirotor aircraft depends primarily on the mission to be performed, which affects the selection of components and parameters of other subsystems. In addition to standard applications where multirotor UAVs are used in data

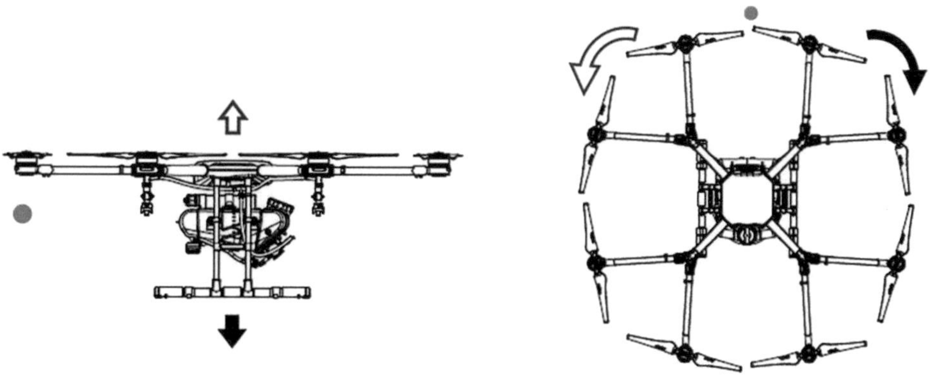

Figure 4.
DJI Agras representation [41].

collection missions, mainly using different types of cameras, they can also be used in special applications. Since the paper considers the application in precision agriculture in smart spraying tasks, the payload of the aircraft is divided into two segments. The first segment consists of the equipment in charge of distributing and spraying the chemical under pressure. The essential parts are a set of hoses and manifolds, sprinklers, nozzles, and pump assembly. It is mounted on the existing aircraft frame, mainly on the landing gear or propulsion arms. The second segment consists of a tank containing a chemical that has a variable mass since it is deployed during the mission.

One of the most widely used commercial aircraft for agricultural purposes is the DJI Agras MG-1, an electric motor multirotor UAV with protection against dust and water. It is designed for applications in a variety of environments and terrains and can be used in fields, terraces, orchards, or other areas. It uses a microwave radar located on the underside of the aircraft that in combination with an altitude stabilization system maintains the aircraft at the desired height above the plants in order to ensure optimal spraying. The volume of the tank is 10 liters, and according to the manufacturer's specifications, it can cover an area of 7–10 acres per hour. The spray mechanism consists of four sprinklers located on two sides of the aircraft. The diameter of the aircraft is 1520 mm, and the configuration consists of eight rotors (octorotor) placed in one plane as shown in **Figure 4** [41].

3.2 Electric energy subsystem

As already mentioned, multirotor UAVs are characterized by high energy consumption as they use rotating wings (propellers) to move in 3D space. The energy subsystem must provide sufficient energy to the aircraft to perform the intended missions and must be compatible with the components of the propulsion subsystem. When selecting the components and parameters of the energy subsystem, the energy requirements of the propulsion subsystem must be taken into account, which in turn depends on the mass and size of the aircraft and the number of propulsion units. The energy subsystem consists of one or more lithium polymer (LiPo) batteries and energy distribution elements. LiPo batteries consist of one or more electrochemical cells in which lithium ions transfer charge between electrodes. They are characterized by high energy density and high discharge rate, which allows higher power and consistent energy flow to the propulsion subsystem. The main parameters of LiPo batteries are their mass, capacity, discharge rate (C), and the number of cells that determine the operating voltage (S).

Battery	Capacity (mAh)	Discharge rate	Mass (g)	Dimension (mm)
Tattu 10000	10000	30 C	2741	182*118*68
Tattu Plus 1.0 16000	16000	15 C	4700	224*163*90
Tattu Plus 1.0 22000	22000	25 C	6058	237*173*116
DJI MG-12000S	12000	20 C	3800	195*151*70

Table 1.
Typical characteristics of high voltage (12S) LiPo batteries [43].

Batteries are the heaviest elements of the aircraft system and have the greatest impact on aircraft dynamics, so it is advisable to place them as close as possible to the aircraft center of gravity. Battery capacity also plays an important role as the flight time of the aircraft depends on it. Hence, the ratio of mass and capacity of the battery is one of the key data when designing a multirotor UAV system. The parameters of the considered Gens ace commercial high-voltage (12S) batteries are listed in **Table 1**. In addition to batteries, the energy subsystem consists of sophisticated circuits for energy distribution and measurement of electrical parameters of the battery.

3.3 Electric propulsion subsystem

The propulsion subsystem of a multirotor UAV is determined by the parameters of the geometric arrangement of the configuration and the characteristics of the propulsion units that make it up. All designs of the propulsion subsystem (configurations) have in common that they consist of N propulsion units (rotors) that generate the necessary forces and moments for the movement of the aircraft in 3D space. Conventional multirotor configurations generally consist of an even number of equal rotors symmetrically arranged in one or more parallel planes. Each pair consists of CW and CCW rotors for the purpose of canceling the reactive moment about the vertical axis of the aircraft. The required performance of the aircraft depends on the type and profile of the mission such as payload, flight duration, power consumption, or other specific requirements. The choice of the propulsion configuration and the type of propulsion units is the key step in the design of the multirotor type of UAV because the flight performance depends on it. **Figure 5** shows the configurations on the same scale of the six-rotor configuration considered in this paper and the eight-rotor configuration that makes up the propulsion subsystem of the DJI Agras commercial aircraft.

The considered electric propulsion units (EPUs) enable precise and fast regulation of control forces and moments that directly affect the position and orientation of the aircraft. The EPU consists of an electronic unit (driver) and a mechanical motor assembly on whose rotor a fixed-pitch propeller is mounted. The brushless DC (BLDC) motor is the central part of the EPU for which there are mostly detailed manufacturer specifications with relevant collocation of driver and propeller. There are EPU components on the market with a very wide choice of motor power, so they can be used in a wide range of multirotor applications, including precision agriculture missions such as smart spraying tasks where carrying a heavier payload is required. The motor speed is controlled by an integrated power inverter, the so-called electronic speed controller (ESC), which generates the switching sequence of the motor phases for the desired RPM specified by the control unit. The rotor of the propulsion unit on which the propeller with fixed pitch is mounted creates aerodynamic forces and moments necessary for the movement of the aircraft. BLDC motor is defined with motor velocity constant

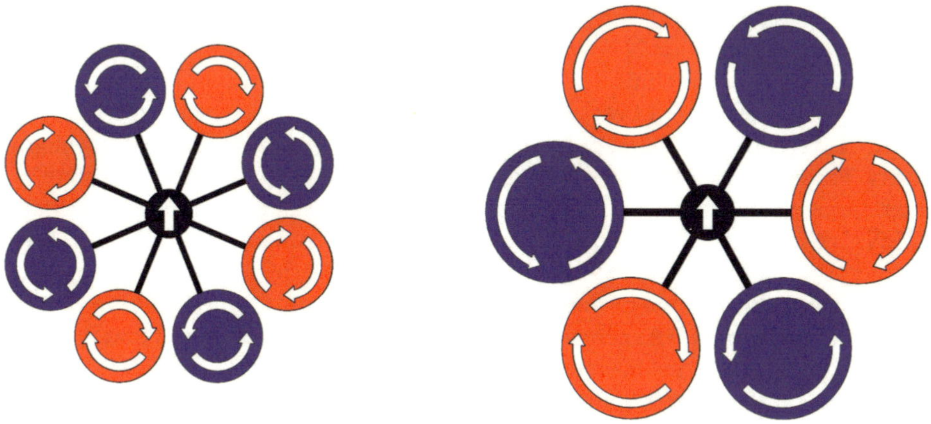

Figure 5.
Conventional multirotor UAV configurations.

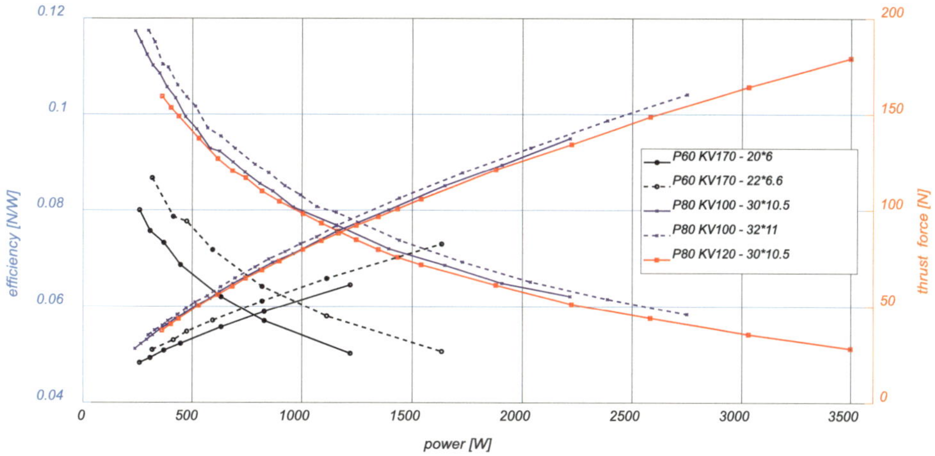

Figure 6.
Considered EPU characteristics [44].

(back EMF constant) Kv. Motors of low power, small dimensions, and large motor constants are used mainly to power micro and small aircraft intended for entertainment or sports (drone racing). On the other hand, high-power and large-dimensions motors with small motor constants are intended for heavy equipment and loads (heavy lift).

In this study, for the needs of the aerial robotic system concept, five combinations of EPUs are considered, which are combined with a high-voltage (12S) energy subsystem setup. Based on the specification of the propulsion components manufacturer, the characterization of EPUs intended for heavy payloads was performed. Selected BLDC motors have a low motor velocity constant (Kv <200), which means that they have lower speeds, so in combination with larger-diameter propellers, they achieve higher torques. **Figure 6** shows the thrust force and efficiency of EPUs as a function of electrical power for the five considered setups. Propeller designations indicate geometry where the first two numbers indicate the propeller diameter in inches, e.g., a propeller marked G32x11 has a diameter of 32″. The next two numbers indicate the pitch of the propeller, also in inches, as the distance that propeller advances during one revolution.

3.4 Control subsystem

The basic task of the control subsystem is to guide the multirotor UAV in 3D space according to the given input variables. In addition, it takes care of the functioning of the entire system and is a kind of interface between the multirotor and the docking facility. The control subsystem primarily consists of a flight controller (FC), state estimation sensors, telemetry, and a remote control receiver. Since the multirotor type of UAVs is characterized by inherent instability, the key component of the aircraft is FC, and it can be freely said that it represents the brain of the aircraft. To control the aircraft concept that would be used in precision agriculture, Pixhawk open-source FC is being considered. The control algorithm generates control signals that it sends to the propulsion units in order to achieve the desired movement in 3D space, i.e., to perform the mission. Orientation sensors are integrated into the Pixhawk FC, and as for the position of the aircraft, it is obtained using a peripheral compatible GPS.

From the aspect of system design, the control subsystem is very demanding because, in addition to the choice of hardware, it is necessary to design a software solution. The considered control unit has already been used in the research so far, and certain segments of code have been tested. **Figure 7** schematically shows the custom firmware that is planned to be used in the future to control the aircraft in precision agriculture.

A series of experiments were conducted to primarily verify the motor mixer subsystem for different aircraft configurations. This will be extremely important for implementation on a prototype aircraft as configurations with different geometric arrangement parameters and with different propulsion unit characteristics have

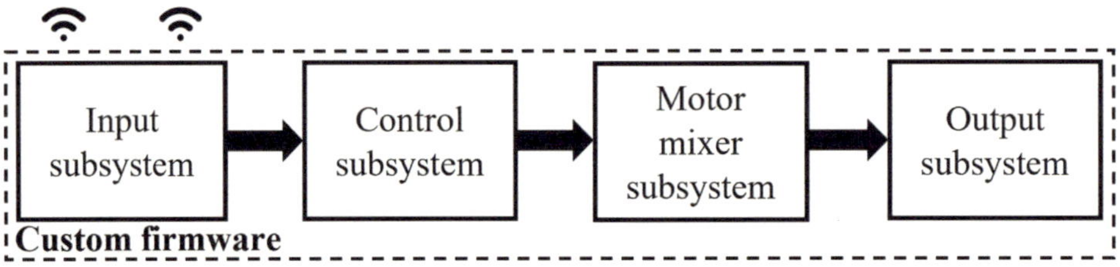

Figure 7.
Schematic representation of custom firmware main subsystems.

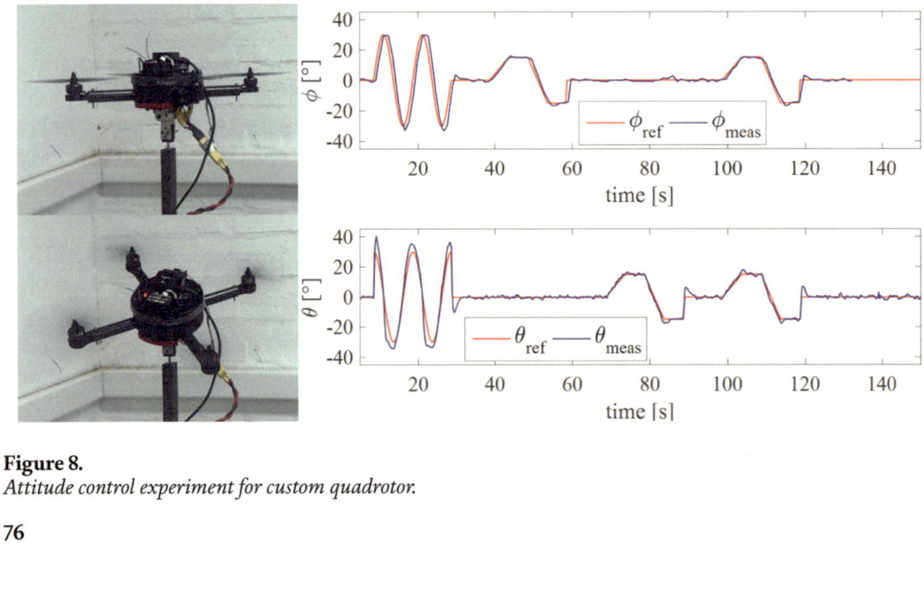

Figure 8.
Attitude control experiment for custom quadrotor.

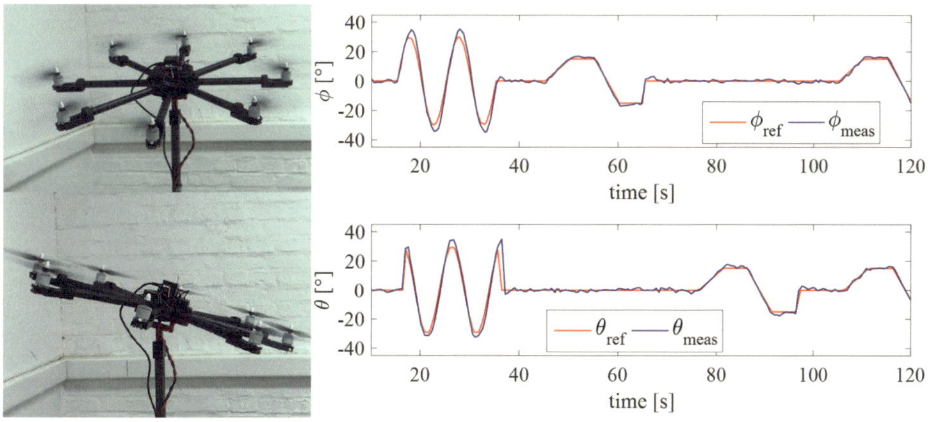

Figure 9.
Attitude control experiment for custom octorotor.

been tested. The first series of experiments was done with a small custom-made quadrotor with x-arrangement. **Figure 8** shows the experimental results of reference attitude tracking. In the next series of experiments, a configuration consisting of eight rotors in a + arrangement, so-called octorotor, was tested (**Figure 9**).

4. Toward heavy-payload multirotor UAV prototype

This chapter will present the results of individual design phases of a multirotor aircraft that is planned to be used as an integral part of the presented concept of an air robotic system for applications in precision agriculture. Experimental measurements of the considered propulsion units were conducted, on the basis of which payload analysis was performed for several configurations. Based on obtained physical parameters, a model was set up and preliminary simulations were performed, with the help of which it is possible to estimate the energy consumption of a real system.

4.1 Electric propulsion unit characterization

In order to adequately select aircraft components to ensure the performance of aircraft required for certain tasks (maximum cargo weight, flight speed, flight time, others), it is important to determine the thrust generated by a specific combination of motor and propeller and to determine power consumption. Based on a certain thrust, the maximum load capacity of the aircraft is determined with regard to the defined thrust-to-weight ratio. Based on electricity consumption, more specifically through the relationship between electric current and thrust, it is possible to estimate the maximum flight time depending on the mission. Manufacturers of propulsion elements generally also provide specifications, as previously shown in **Figure 6**; however, these data are not in all cases consistent with actual characteristics. Therefore, for a more precise analysis of the propulsion, it is necessary to perform characterization, and in this paper, the method described in the previous research was used [45] utilizing the experimental test stand RCbenchmark 1780 [46]. **Figure 10** shows the thrust force as a function of the angular velocity of the rotor for the considered propulsion units where the measured experimental characteristics and the characteristics according

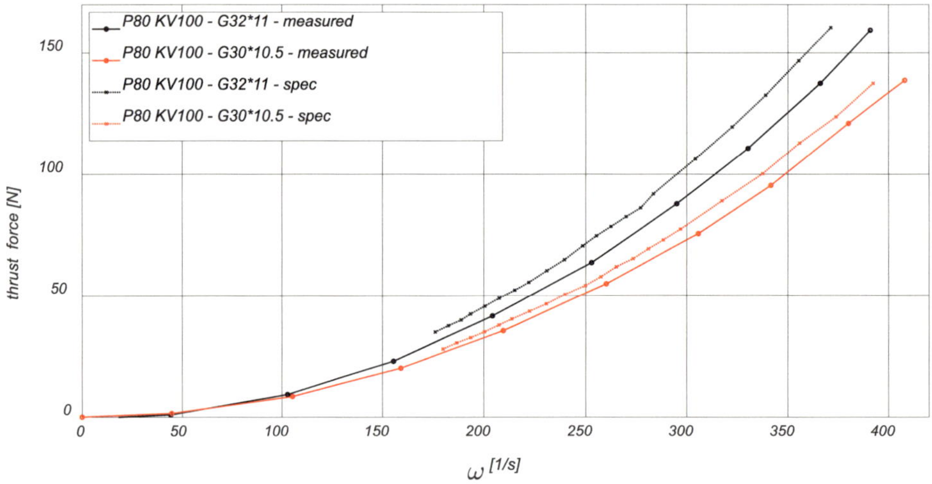

Figure 10.
Thrust force with respect to rotor angular velocity.

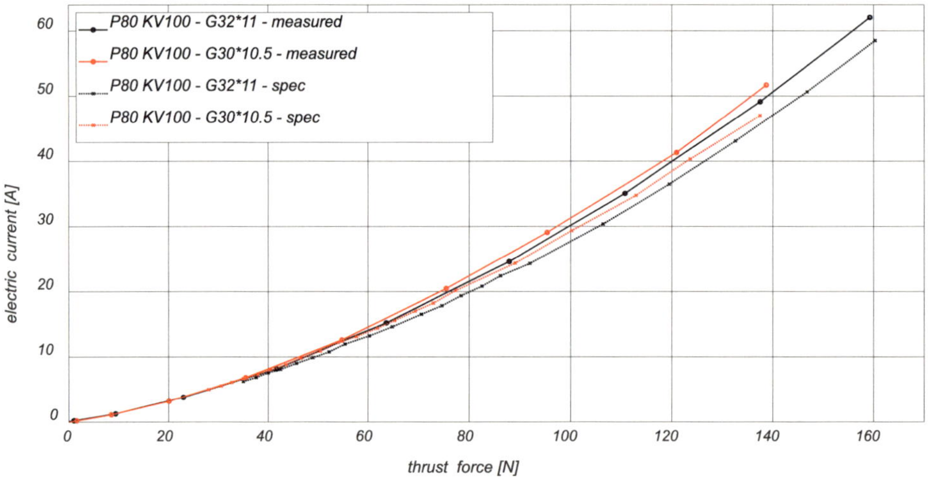

Figure 11.
Electric current with respect to the thrust force.

to the manufacturer's specifications are shown. Furthermore, **Figure 11** shows the electric current as a function of the thrust force for the purpose of estimating the flight time.

4.2 System mass distribution analysis

As mentioned in the previous sections, the weight (mass) of the aircraft plays an important role as it will directly affect the maximum payload of the aircraft. In order to be able to accurately determine the payload of an aircraft, the weight of all aircraft components/subsystems has to be known. Taking into account the choice of propulsion components, and the configuration of the aircraft, the choice of the energy subsystem will greatly affect the carrying capacity of the aircraft. **Figure 12** graphically

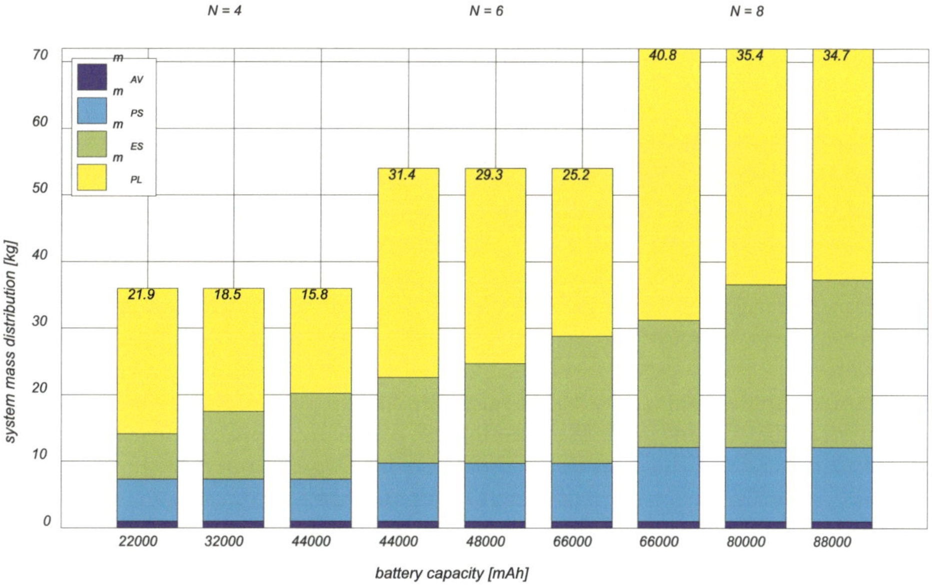

Figure 12.
System mass distribution for three conventional configurations.

shows the dependence of the mass distribution of the aircraft subsystems in the case of three conventional aircraft configurations and various battery capacities. It can be seen that the mass of the avionics (control) subsystem can be considered fixed since the components that make up the control subsystem do not change in relation to the changes of other subsystems. The mass of the propulsion subsystem varies with the number of EPUs required to perform certain missions and significantly affects the total mass of the system. In terms of energy consumption, more units will require more energy, which means that more batteries will be needed, and the mass of the batteries, i.e., the mass of the energy subsystem, has the greatest impact on the total mass. All this affects the maximum payload of the aircraft. A larger number of EPUs will generally provide higher thrust and a higher payload mass, although they will also require a heavier energy subsystem with the ability to deliver more energy. The process of designing a multirotor aircraft is extremely demanding, especially given the limitations that exist in the size of the aircraft, but also energy consumption (**Figure 12**).

Although a change in battery capacity will not change the overall thrust generated by the propulsion subsystem, it will affect the overall mass of the system and thus the payload of the aircraft and the flight time. The higher-capacity batteries have an expected higher mass, thus leaving less space for payload mass and requiring higher energy consumption to compensate for heavier aircraft. Thus, a higher-capacity battery does not always result in a longer flight time.

Since the system is divided into four key subsystems, as mentioned earlier, a certain degree of modularity is allowed. In the further work, special attention will be paid in the design phase to the construction of modular elements, which would allow easy assembly of aircraft configurations with different numbers of rotor arms, thus further expanding the diversity of the system and potentially reducing energy consumption. In this sense, the guidelines presented in the previous work [47] regarding the small educational aircraft will be used.

4.3 Simulation results

In the use of UAV for spraying or similar tasks such as fertilization or even seed sowing, the payload capacity is specific. As the aircraft tank is filled with the required chemicals (either fertilizer or seed) and depleted during usage, the weight (mass) of the aircraft will also continuously decrease. In order to efficiently conduct the spraying task with low energy and time losses, the flight path needs to be planned with regard to the tank size and the chemical consumption rate. The rate of chemical consumption is also not fixed for the whole parcel but depends on the crop health condition estimated based on sensor readings. Flight planning is an extremely complex process that includes many parameters, which will be the subject of future research.

To determine the energy consumption of the aircraft during the spraying mission and to approximately determine the required flight time, it is necessary to conduct computer simulations in the development phase of the prototype. In this way, the development time and the price of the product can be significantly shortened, as the possibility of incorrect selection of system components and parameters is reduced. Preliminary simulations are presented in this paper, where typical spraying parameters are taken: nozzle spraying rate of 0.375 L/min, spray width of 5 m, and flying speed of 2 m/s. The aircraft is equipped with a spraying tank of 25 L volume, and four spraying nozzles, which gives a total spraying rate of 1.5 L/min. Based on those specifications, a minimum flight time of 16.5 min is required to deplete the whole tank, and in that time area of approximately 10000 m^2 can be covered. The aircraft parameters (mass and inertia) were obtained based on a simplified 3D CAD model. **Figure 13** shows the most elementary case when the mission consists of uniform spraying of the crop. Air resistance or any disturbances are not included in the simulations, this is planned in the next phases of the research.

Based on the planned flight consisting of take-off, horizontal flight in the pattern, and landing, the angular velocities of individual EPUs or direct control signals (PWM) can be extracted from the model, as shown in **Figure 14**. As mentioned, with the consumption of the chemical, the mass of the aircraft is reduced, which results in fewer forces and moments of the propulsion subsystem required for motion in 3D space, which can be seen in the figure where the control signals are continuously reduced. The main goal of the simulation is to determine the energy consumption of the aircraft by approximating the individual energy consumption of each EPU, which can be determined if the flight pattern and the change in aircraft mass are known. Since

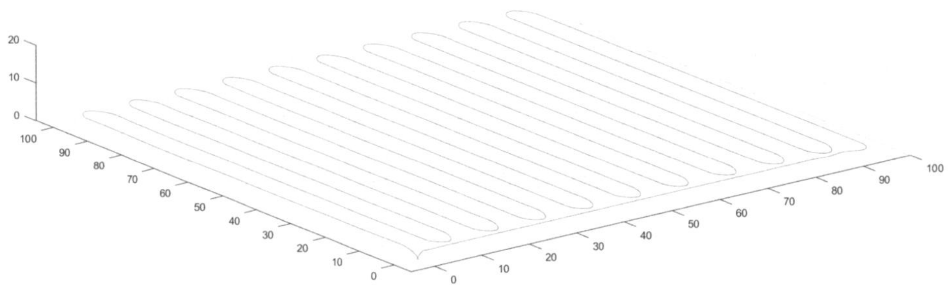

Figure 13.
An example of the aircraft trajectory in a spraying mission.

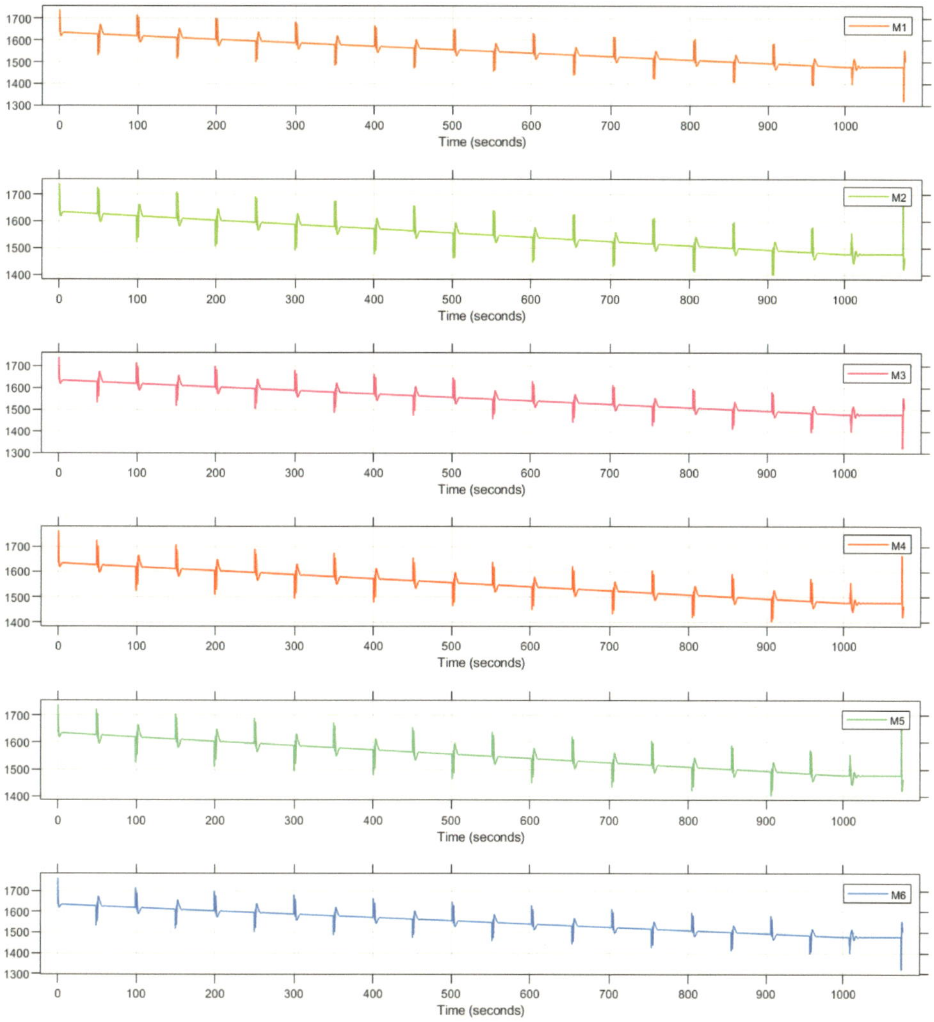

Figure 14.
Motor control signals related to given spraying mission.

there are characteristics of propulsion units, it is easy to connect electrical quantities (electric current, voltage, and electric power) with the control signal or the angular velocity of the rotor. This can further allow the selection of optimal system components and parameters, which is extremely important in the system design phase.

5. Conclusion

This paper discusses the current state of the art regarding the use of multirotor UAVs for spraying tasks in precise agriculture. The possibilities of application of the proposed autonomous aerial robotic system consisting of a mobile base station and a multirotor type of UAV were demonstrated. The purpose of the presented system was to autonomously perform spraying tasks on different ranges of surfaces, including large crops parcels. In such a system, special emphasis was placed on the functions of the mobile base station, which had to provide support for autonomous spraying and be a user interface. By selecting the correct components and parameters of the aircraft system, satisfactory spraying coverage, flight performance, and flight duration were achieved. In the future work, it is planned in the first phase to prototype the aircraft and then extensive testing of the control module. In the second phase, it is planned to design custom aircraft equipment and a mobile base station.

Acknowledgements

This research was funded by European Regional Development Fund, Operational programme competitiveness and cohesion 2014–2020, as part of the call for proposals entitled “Investing in science and innovation—first call,” grant number KK.01.1.1.04.0092.

Author details

Petar Piljek[1], Marko Pranjić[2], Denis Kotarski[2*] and Tomislav Petanjek[3]

1 Faculty of Mechanical Engineering and Naval Architecture, University of Zagreb, Zagreb, Croatia

2 Karlovac University of Applied Sciences, Karlovac, Croatia

3 DOBA Business School, Maribor, Slovenia

*Address all correspondence to: denis.kotarski@vuka.hr

References

[1] Cisternas I, Velásquez I, Caro A, Rodríguez A. Systematic literature review of implementations of precision agriculture. Computers and Electronics in Agriculture. 2020;**176**:105626. DOI: 10.1016/j.compag.2020.105626

[2] Shafi U, Mumtaz R, Garcia-Nieto J, Ali Hassan S, Ali Raza Zaidi S, Iqbal N. Precision agriculture techniques and practices: From considerations to applications. Sensors. 2019;**19**:3796

[3] Sung J. The fourth industrial revolution and precision agriculture. In: Hussmann S, editor. Automation in Agriculture - Securing Food Supplies for Future Generations. London: IntechOpen; 2018. DOI: 10.5772/intechopen.71582

[4] Radoglou-Grammatikis P, Sarigiannidis P, Lagkas T, Moscholios I. A compilation of UAV applications for precision agriculture. Computer Networks. 2020;**172**:107148. DOI: 10.1016/j.comnet.2020.107148

[5] Zhang C, Kovacs JM. The application of small unmanned aerial systems for precision agriculture: A review. Precision Agriculture. 2012;**13**:693-712. DOI: 10.1007/s11119-012-9274-5

[6] Virnodkar SS, Pachghare VK, Patil VC, Kumar JS. Remote sensing and machine learning for crop water stress determination in various crops: A critical review. Precision Agriculture. 2020;**21**:1121-1155. DOI: 10.1007/s11119-020-09711-9

[7] Adamchuk VI, Viscarra Rossel RA, Sudduth KA, Lammers PS. Sensor fusion for precision agriculture. In: Thomas C, editor. Sensor Fusion: Foundation and Applications. London: IntechOpen; 2011. DOI: 10.5772/19983

[8] Saleem MH, Potgieter J, Arif KM. Automation in agriculture by machine and deep learning techniques: A review of recent developments. Precision Agriculture. 2021;**22**:2053-2091. DOI: 10.1007/s11119-021-09806-x

[9] Zhang C, Valente J, Kooistra L, Guo L, Wang W. Orchard management with small unmanned aerial vehicles: A survey of sensing and analysis approaches. Precision Agriculture. 2021;**22**:2007-2052. DOI: 10.1007/s11119-021-09813-y

[10] Hunt ER Jr, Horneck DA, Spinelli C, Turner RW, Bruce AE, Gadler DJ, et al. Monitoring nitrogen status of potatoes using small unmanned aerial vehicles. Precision Agriculture. 2018;**19**:314-333. DOI: 10.1007/s11119-017-9518-5

[11] Rasmussen J, Nielsen J, Streibig JC, Jensen JE, Pedersen KS, Olsen SI. Pre-harvest weed mapping of Cirsium arvense in wheat and barley with off-the-shelf UAVs. Precision Agriculture. 2019;**20**:983-999. DOI: 10.1007/s11119-018-09625-7

[12] Rahim H, Ghazali MS, Bookeri MA,·Bakar BH, Ariff E, Rahman M, Wahab MA. Economic potential of rice precision farming in malaysia: The case study of Felcra Seberang Perak. Precision Agriculture. 2021

[13] Young SN, Kayacan E, Peschel JM. Design and field evaluation of a ground robot for high-throughput phenotyping of energy sorghum. Precision Agriculture. 2019;**20**:697-722. DOI: 10.1007/s11119-018-9601-6

[14] Gonzalez-de-Soto M, Emmi L, Perez-Ruiz M, Aguera J, Gonzalez-de-Santos P. Autonomous systems

for precise spraying e Evaluation of a robotised patch sprayer. Biosystems Engineering. 2016;**146**:165-182. DOI: 10.1016/j.biosystemseng.2015.12.018

[15] Cantelli L, Bonaccorso F, Long D, Melita CD, Schillaci G, Muscato G. A small versatile electrical robot for autonomous spraying in agriculture. AgriEngineering. 2019;**1**:391-402. DOI: 10.3390/agriengineering1030029

[16] Quaglia G, Visconte C, Sabatino Scimmi L, Melchiorre M, Cavallone P, Pastorelli S. Design of a UGV powered by solar energy for precision agriculture. Robotics. 2020;**9**:13. DOI: 10.3390/robotics9010013

[17] Gorjian S, Minaei S, MalehMirchegini L, Trommsdorff M, Shamshiri RR. Applications of solar PV systems in agricultural automation and robotics. In: Gorjian S, Shukla A, editors. Photovoltaic Solar Energy Conversion. Cambridge, Massachusetts: Academic Press; 2020. pp. 191-235. DOI: 10.1016/B978-0-12-819610-6.00007-7

[18] Tokekar P, Vander Hook J, Mulla D, Isler V. Sensor planning for a symbiotic UAV and UGV system for precision agriculture. In: Proceedings of the IEEE International Conference on Intelligent Robots and Systems (IROS '13); 3-7 November 2013; Tokyo. pp. 5321-5326. DOI: 10.1109/IROS.2013.6697126

[19] Vu Q , Raković M, Delić V, Ronzhin A. Trends in development of UAV-UGV cooperation approaches in precision agriculture. In: Proceedings of the Interactive Collaborative Robotics (ICR'18); 18-22 September 2018; Leipzig. New York: Springer; 2018 pp. 213-221. DOI: 10.1007/978-3-319-99582-3_22

[20] Castaldi F, Pelosi F, Pascucci S, Casa R. Assessing the potential of images from unmanned aerial vehicles (UAV) to support herbicide patch spraying in maize. Precision Agriculture. 2017;**18**:76-94

[21] Gonzalez-de-Santos P, Ribeiro A, Fernandez-Quintanilla C, Lopez-Granados F, Brandstoetter M, Tomic S, et al. Fleets of robots for environmentally-safe pest control in agriculture. Precision Agriculture. 2017;**18**:574-614. DOI: 10.1007/s11119-016-9476-3

[22] Yang MD, Boubin JG, Tsai HP, Tseng HH, Hsu YC, Stewart CC. Adaptive autonomous UAV scouting for rice lodging assessment using edge computing with deep learning EDANet. Computers and Electronics in Agriculture. 2020;**179**:105817

[23] Shamshiri RR, Ibrahim AH, Siva KB, Desa A, Cornelia W, Muhammad Y. Fundamental research on unmanned aerial vehicles to support precision agriculture in oil palm plantations. In: Zhou J, editor. Agricultural Robots - Fundamentals and Application. London: IntechOpen; 2018. pp. 91-116. DOI: 10.5772/intechopen.80936

[24] Rokhmana CA. The potential of UAV-based remote sensing for supporting precision agriculture in Indonesia. Procedia Environmental Sciences. 2015;**24**:245-253. DOI: 10.1016/j.proenv.2015.03.032

[25] Li M, Shamshiri RR, Schirrmann M, Weltzien C, Shafian S, Laursen MS. UAV oblique imagery with an adaptive micro-terrain model for estimation of leaf area index and height of maize canopy from 3D point clouds. Remote Sensing. 2022;**14**(3):585. DOI: 10.3390/rs14030585

[26] Barbedo JG. A review on the use of unmanned aerial vehicles and imaging sensors for monitoring and assessing plant stresses. Drones. 2019;**3**(2):40. DOI: 10.3390/drones3020040

[27] Wang L, Yubin L, Xuejun Y, Kangjie L, Zhenzhao C, Ziyao C, et al. Vision-based adaptive variable rate spraying approach for unmanned aerial vehicles. International Journal of Agricultural and Biological Engineering. 2019;**12**(3):18-26

[28] Lian Q, Tan F, Fu X, Zhang P, Liu X, Zhang W. Design of precision variable-rate spray system for unmanned aerial vehicle using automatic control method. International Journal of Agricultural and Biological Engineering. 2019;**12**(2):29-35

[29] Zhang Y, Huang X, Lan Y, Wang L, Lu Y, Yan K, et al. Development and prospect of UAV-based aerial electrostatic spray technology in China. Applied Sciences. 2021;**11**:4071. DOI: 10.3390/app11094071

[30] Xue X, Lan Y, Sun Z, Chang C, Hoffmann WC. Develop an unmanned aerial vehicle based automatic aerial spraying system. Computers and Electronics in Agriculture. 2016;**128**:58-66. DOI: 10.1016/j.compag.2016.07.022

[31] Martinez-Guanter J, Agüera P, Agüera J, Pérez-Ruiz M. Spray and economics assessment of a UAV-based ultra-low-volume application in olive and citrus orchards. Precision Agriculture. 2020;**21**:226-243. DOI: 10.1007/s11119-019-09665-7

[32] Basso M, Stocchero D, Ventura Bayan Henriques R, Vian AL, Bredemeier C, Aparecida Konzen A, et al. Proposal for an embedded system architecture using a GNDVI algorithm to support UAV-based agrochemical spraying. Sensors. 2019;**19**:5397

[33] Ivić S, Andrejčuk A, Družeta S. Autonomous control for multi-agent non-uniform spraying. Applied Soft Computing. 2019;**80**:742-760. DOI: 10.1016/j.asoc.2019.05.001

[34] Faiçal BS, Freitas H, Gomes PH, Mano LY, Pessin G, de Carvalho ACPLF, et al. An adaptive approach for UAV-based pesticide spraying in dynamic environments. Computers and Electronics in Agriculture. 2017;**138**:210-223. DOI: 10.1016/j.compag.2017.04.011

[35] Faiçal BS, Costa FG, Pessin G, Ueyama J, Freitas H, Colombo A, et al. The use of unmanned aerial vehicles and wireless sensor networks for spraying pesticides. Journal of Systems Architecture. 2014;**60**:393-404. DOI: 10.1016/j.sysarc.2014.01.004

[36] Yang F, Xue X, Cai C, Sun Z, Zhou Q. Numerical simulation and analysis on spray drift movement of multirotor plant protection unmanned aerial vehicle. Energies. 2018;**11**:2399. DOI: 10.3390/en11092399

[37] Wen S, Zhang Q, Deng J, Lan Y, Yin X, Shan J. Design and experiment of a variable spray system for unmanned aerial vehicles based on PID and PWM control. Applied Sciences. 2018;**8**:2482. DOI: 10.3390/app8122482

[38] Kharima MNA, Wayayoka KA, Shariff ARM, Abdullah AF, Husin EM. Droplet deposition density of organic liquid fertilizer at low altitude UAV aerial spraying in rice cultivation. Computers and Electronics in Agriculture. 2019;**167**:105045. DOI: 10.1016/j.compag.2019.105045

[39] Wang G, Lan Y, Yuan H, Qi H, Chen P, Ouyang F, et al. Comparison of spray deposition, control efficacy on wheat aphids andworking efficiency in the wheat field of the unmanned aerial vehicle with boom sprayer and two conventional Knapsack Sprayers. Applied Sciences. 2019;**9**:218. DOI: 10.3390/app9020218

[40] Meng Y, Su J, Song J, Chen WH, Lan Y. Experimental evaluation of UAV

spraying for peach trees of different shapes: Effects of operational parameters on droplet distribution. Computers and Electronics in Agriculture. 2020;**170**:105228. DOI: 10.1016/j.compag.2020.105282

[41] Agras MG-1 User Manual v1.2. Available from: https://dl.djicdn.com/downloads/mg-1/en/MG-1_User_Manual_en_v1.2.pdf [Accessed: January 5, 2022]

[42] DJI Agras MG-1 Commercial Aircraft. Available from: https://www.dji.com/hr/mg-1s [Accessed: January 7, 2022]

[43] LiPo batteries Gens Ace. Available from: https://www.gensace.de/lipo-12s [Accessed: January 7, 2022]

[44] Electric Propulsion Tiger motor P type. Available from: https://store.tmotor.com/category.php?id=42 [Accessed: January 5, 2022]

[45] Piljek P, Kotarski D, Krznar M. Method for characterization of a multirotor UAV electric propulsion system. Applied Sciences. 2020;**10**:8229. DOI: 10.3390/app10228229

[46] RCbenchmark Series 1780 25 & 40 kg Test Stand User Manual. Available from: https://cdn-docs.rcbenchmark.com/manuals/series-1780/Series%20 1780%20v2.5%20General%20User%20 Manual.pdf [Accessed: December 22, 2022]

[47] Kotarski D, Piljek P, Pranjić M, Grlj CG, Kasać J. A modular multirotor unmanned aerial vehicle design approach for development of an engineering education platform. Sensors. 2021;**21**:2737. DOI: 10.3390/s21082737

Chapter 4

Digital Agriculture in Iran: Use Cases, Opportunities, and Challenges

Seyed Moin-eddin Rezvani, Redmond R. Shamshiri, Jalal Javadi Moghaddam, Siva K. Balasundram and Ibrahim A. Hameed

Abstract

Agriculture is constantly developing into a progressive sector by benefiting from a variety of high-tech solutions with the ultimate objectives of improving yield and quality, minimizing wastes and inputs, and maximizing the sustainability of the process. For the case of Iran, adaptation of digital agriculture is one of the key economic plans of the government until 2025. For this purpose, the development of infrastructure besides understanding social and cultural impacts on the transformation of traditional agriculture is necessary. This chapter reports the potential of the existing technological advances and the state of the current research efforts for the implementation of digital agriculture in open-field and closed-field crop production systems in Iran. The focus of the study was on the development of affordable IoT devices and their limitations for various farming applications including smart irrigations and crop monitoring, as well as an outlook for the use of robotics and drone technology by local farmers in Iran.

Keywords: digital economy, greenhouse, irrigation, robotic, smart, intelligent

1. Introduction

Deficiency of water resources and arable land along with global climate change are the main limiting factors for feeding the growing population in the world. The per capita arable land worldwide from 1961 to 2018 decreased from 0.361 hectares to 0.184 hectares (97% reduction), and in Iran, the per capita arable land decreased from 0.666 to 0.179 hectares (272% reduction). The per-person renewable water in the world from 1962 to 2017 decreased from 13,407 to 5724 cubic meters (134% reduction) and in Iran from 5570 to 1593 cubic meters (250% reduction) [1]. According to the FAO, the world's population will reach 10 billion by 2050, and with moderate economic growth, the need for food will increase by 50% compared to 2013. The scarcity of production resources and reducing environmental impacts have necessitated the need to increase the productivity of the resources.

According to UNCTAD's 2019 report, the share of digital economy in relation to Iran's GDP rose from 2.2% in 2012 to 6.5% in 2020. Precision agriculture makes it possible to increase the productivity of production factors and reduce the environmental risks. As defined by the International Association for Precision Agriculture [2]: *"Precision Agriculture is a management strategy that gathers, processes and analyzes temporal, spatial and individual data and combines it with other information to support management decisions according to estimated variability for improved resource use efficiency, productivity, quality, profitability and sustainability of agricultural production."* The evolution of precision agriculture has been made possible through the automatic collection, integration, and analysis of data silos previously isolated from the field, equipment sensors, and other third-party sources, using Industry 4.0 intelligent and digital technologies, leading to Agriculture 4.0 (or Digital Agricultural) [3]. From 12,000 years ago, when the agricultural revolution led to the settlement and the emergence of civilizations, to about one hundred years ago that the agricultural mechanization revolution took place, the changes were slow. The use of modified and agrochemical products developed in the 1960s, which was completed with the advent of genetic technology in the last decade of the past century [4]. Digital agriculture by Internet of Things (IoT), cloud computing, and Big data analysis collected and analyzed the required data from the farm by sensing, data management, data processing, and data enhancement. The analyzed results for decision making or activation were provided to farmers, agricultural robots, automation, or decision support systems [5–7]. The digital agricultural revolution will change not only farm operations but also every part of the value chain of agricultural products [4]. Digital agriculture has provided the possibility of generating knowledge to support the farmer in the decision-making process in the farm enterprise.

Digital agriculture brings the possibility of higher output with lower input resources by providing tools and methods for measuring the environment, processing information and accurate operations in combination with an integrated digital system with market status information, communication between stakeholders, interaction with buyers of products, and agricultural service providers giving the farmer ability to get the most out of the market [5]. Based on wireless sensor, and positioning technologies, data analysis solutions, mobile applications, and web-based solutions, the main technologies used in digital agriculture are sensor-based field mapping, wireless crop monitoring, climate monitoring and forecasting, stats on-farm production, monitor wireless equipment, predictive analytics for crop and livestock, livestock tracking and geo-referencing, and smart logistics and warehousing [5, 8, 9]. Salam A. [10] studied the barriers to the acceptance of digital agriculture found out the main obstacle is the return on investment. The next hurdle is the lack of attention to small farm owners in the digital technology business and the focus on large farms. In addition to the diversity of digital farming technologies in the fields of topography and soil texture and the lack of decision tools for the enormous data being generated from the farm, decision-making is very time consuming for farmers. They prefer to make decisions based on their experience. Other barriers to accurate trade are cost and the availability of specialists for complex equipment, lack of manufacturer support, difficulty in putting up encompassing high value, and precision portfolios. Because of these barriers, the digital farming business is not profitable. Da Silveira et al. [11] identified 25 barriers to the development of agriculture 4.0 and, in order of importance classified them into five dimensions: technological, social, political, economic, and environmental, respectively. A review of articles on barriers to the development of agriculture 4.0 showed that the key issues were incompatibilities between technological components, concerns about issues of reliability, technological complexity, lack of infrastructure, lack of R&D and innovative business models, lack of digital skills or skilled labor, information asymmetry between agricultural production chain actors, and problems in education. Less important barriers included sustainable constraints, concerns about environmental, ethical, and social costs, interruption of existing work, age group risks, and concerns about sustainable energy sources.

According to FAO and ITU (International Telecommunication Union), some of the potential risks and barriers to e-agriculture are poor ICT (Information and Communications Technology) and e-agriculture infrastructure; accessibility and inclusivity problems due to inappropriate ICT distribution; marginalization of women in the use of ICT in agriculture; a lack of an inclusive approach with ICTs—attention to differently abled, semiliterate/illiterate users; low levels of e-agriculture best practices, customization, and personalization; high cost of e-agriculture services and the absence of sustainable business models; and the decline of public expenditure on agriculture in developing countries [12]. Bagheri and Kafashian [13] considered the challenges of precision agriculture in Iran as the smallholder and the poor financial strength of most farmers, lack of accurate information on profitability due to the use of precision agricultural technologies, low tendency of mechanization levels, lack of required facilities and equipment, lack of precision agriculture infrastructure, poor knowledge of farmers and executives in the field of precision agriculture, and lack of skilled workforce to train, use, repair and maintain equipment related to precision agriculture. The results of economic analysis based on national statistics and research conducted in Iran show that the application of precision agriculture in the current agricultural conditions reduces costs by 15–40%.

Today, with the increase of the world population, water shortage, energy, arable land, and the need to provide food, traditional agricultural methods no longer meet the food needs of the world population, and the smart farming strategy has received much attention [7, 8, 14–20]. Low productivity of the agricultural sector and limited production resources, especially water, have paved the way for the transformation of the agricultural sector with the help of digital technology in Iran. Optimal use of soil and water resources and other agricultural inputs with increasing productivity and performance is one of the most important advantages of using digital farming systems. Traditional agriculture is becoming more accurate and digital, and Iran will have to adapt to the global agricultural system. The purpose of this chapter is to study the infrastructure and current situation of some digital agriculture aspects in Iran.

2. Digital economy

Due to the high speed of technological change and its use by companies and consumers [21], the definition of the digital economy has evolved over time [22]. Digital economy, according to the definition of the Organization for Economic Co-operation and Development (OECD), is an economy most of which is based on digital technologies, including communication networks, computers, software, and other information technologies, and various types of e-commerce, e-markets. It also includes smart cards, e-money, and financial transactions. According to the UNCTAD (United Nations Conference on Trade and Development) definition, digital economy means the use of Internet-based digital technologies to produce and trade goods and services [21]. Bukht and Heeks [22] defined the digital economy as "that part of economic output derived solely or primarily from digital technologies with a business model based on digital goods or services." Through this approach, the digital economy consists of three layers:

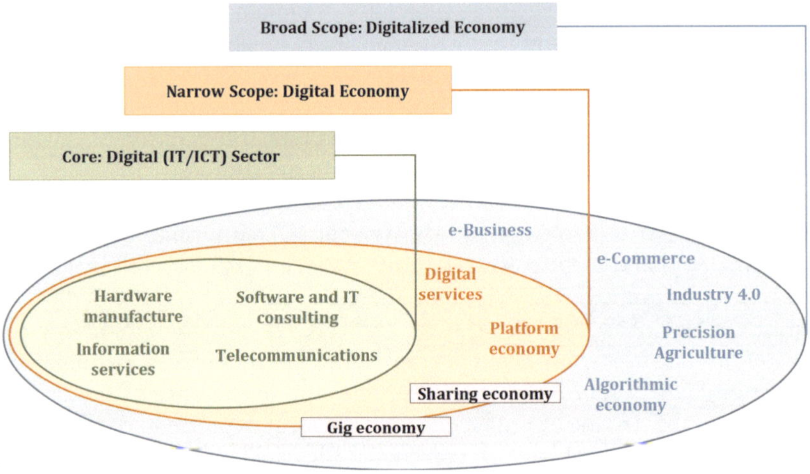

Figure 1.
Illustration of the scopes of the digital economy [22].

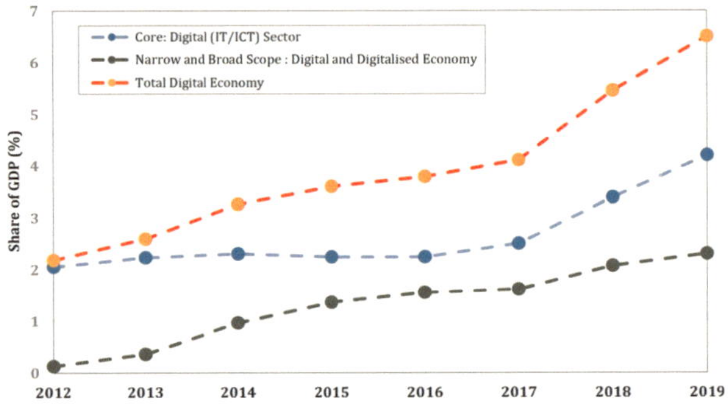

Figure 2.
Changes in digital economy share of GDP in Iran from 2012 to 2019 [23].

first, a core including hardware manufacturing, software, and digital (IT/ICT) sector, the second layer narrow scope including electronic business, digital services, and platform economy (digital economy), and the third layer broad scope including e-commerce and algorithm economy (digitalized economy) (**Figure 1**).

The digital economy share of GDP in Iran increased from 2.2% in 2012 to 6.5% in 2019 (**Figure 2**). Although the core layer with 4% is close to the global average of 4.5%, the second and third layer with 2.5% is still significantly different from the global average of 15.5%. The digital economy in Iran, however, is rapidly growing. According to Tufts University, Iran ranks sixth among the 90 countries surveyed in the world in the momentum (growth rate) of the digital economy [23].

2.1 Digital infrastructure

While the penetration rate of fixed telephones from 2013 to September 2021 shows a decrease of 2.4 percent (**Figure 3a**) and in the years 2006 to September 2021, the

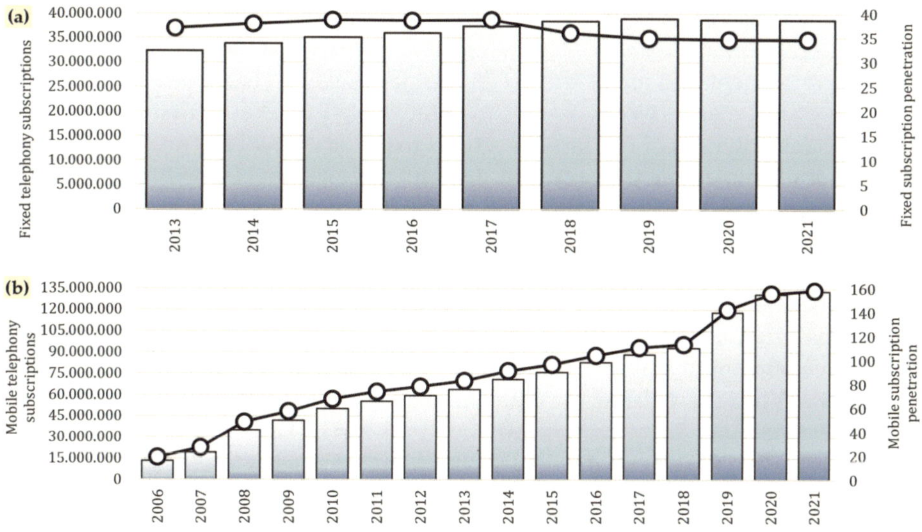

Figure 3.
A comparison between the number of landline users (top) and mobile phone users between 2006 and 2021 in Iran.

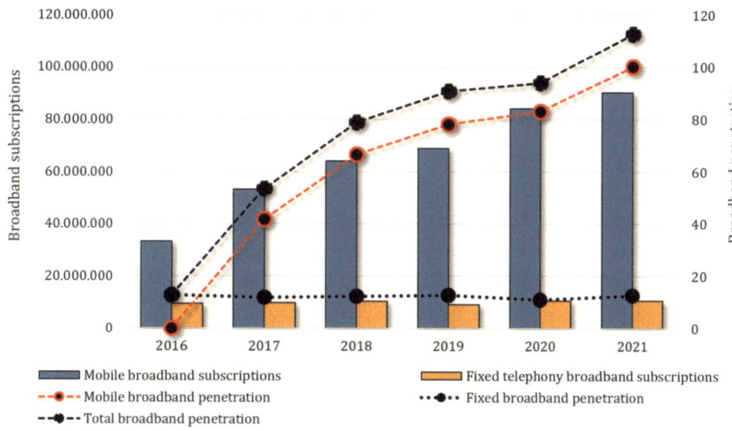

Figure 4.
Fixed telephony and mobile broadband subscriptions and penetration from 2016 to 2021.

penetration rate of mobile phones increased from 18.7 to 154.8 percent (**Figure 3b**). The penetration rate of mobile phones from 2006 to 2017 and 2017 to September 2021 increased by 7.9% and 13.8% every year, respectively [24].

From 2016 to September 2021, the mobile broadband penetration rate reached from zero to 100.22%, while it was around 12.2% for fixed telephone bandwidth (**Figure 4**) [24]. In 2020, the population covered by at least a 3G and 4G mobile network was 85% and 81%, respectively. In 2017, households with Internet access at rural and urban home were 56.98% and 77.92%, respectively, and households with Internet access at home reached 93.30% in 2020. In 2017, individuals with basic, standard, and advanced ICT (Information and Communications Technology) skills were 20.58%, 7.98%, and 1.28%, respectively [24].

2.2 Challenges of IoT in Iran

The development and use of the IoT in Iran face several challenges, and the reluctance of internet service providers to enter the IoT market in Iran [25] have caused the development of IoT infrastructures to be very slow [26], and face shortages [25–27]. For example, due to the lack of demand and the market, operators are reluctant to construct infrastructure, and the prospect of moving to 5G is challenged [25]. One of the most important platforms of the IoT is the migration from the fourth generation of IP addresses to the sixth generation, and it is not clear what the stage is in Iran [26]. The ignorance of various institutions about the powerful applications of IoT, such as smart making and Industrial IoT, is one of the obstacles to the productivity and development of this technology, and consequently, the IoT market prosperity in Iran [28]. While the pillar of IoT implementation is equipping devices with sensors and hardware components that transmit data to the IoT platform, given the current economic situation in the country, the production of these parts or their import has problems, and estimates show that the cost of the existing parts is very high [29]. Another major challenge is the lack of high-performance software platforms for sensor data collection, storage, processing, and analysis, in a short time. Almost none of the powerful foreign platforms inside Iran provide services [29]. Another obstacle is the lack of public awareness of the use of the Internet of Things on a large scale [27, 29]. For the IoT field, there is a need for access to data and measurements (data transparency and open data), but for various reasons, there are problems in the IoT field in Iran [25]. There are also challenges to data transferring to the network for use in Iran. In IoT technology, Zigbee, BLE 5.0, or Wi-Fi can be used to connect devices in the environment to a network that requires short-range connections. While Wi-Fi is present in almost all public and private places, it takes a lot of energy to connect to the network and reduces battery life. Zigbee requires less cost and energy consumption but has a low data transfer rate and is also supported by limited modules in Iran (despite their very high price). There are also standards for long-distance connections such as SigFox, NB-IoT, and LoRa. Despite the high data transmission security by LoRa, it is almost not used in Iran. SigFox protocol was recently launched, and there is still the problem of supporting it in different country regions and devices that can communicate by this protocol. NB-IoT not only has low security in data transmission but also has been piloted by mobile operators and has limited support [29].

Orandi [28] summarized IoT challenges in Iran as follows: (1) The provision of the necessary technical infrastructure has been challenged by international sanctions, (2) there is no proper standardization for smart advice, (3) the lack of separation of smart goods from non-smart goods by customs has created many problems for actors in this field, (4) government institutions and organizations do not function in an integrated way to develop the Internet of Things, (5) the rights and ownership of data collected in IoT are not specified in the country, (6) the role of private sector investment and participation in large national projects is very small, (7) cumbersome rules are in some cases an barrier for IoT developers, and (8) the very important role of universities and research centers in the development of IoT technology has not been considered.

3. Digital agriculture in Iran

Iranian agriculture is in the second stage of the agricultural revolution and is transitioning from the second to third generation agriculture. In recent years,

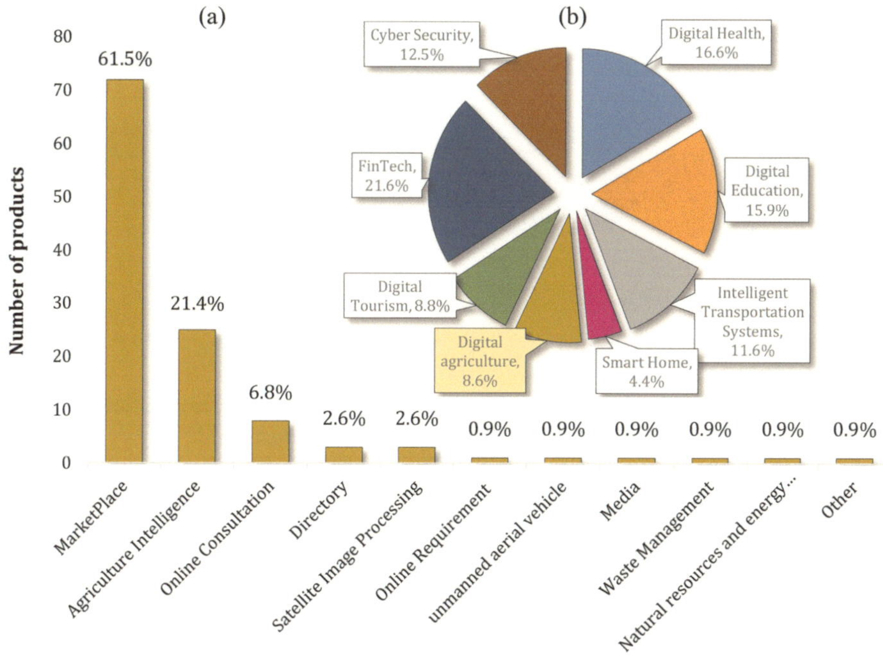

Figure 5.
The share of different sectors of the digital economy (a) and digital agriculture (b).

extensive efforts have been made to develop the technologies of the fourth generation of the industrial revolution in the agricultural sector. All products produced in the third layer of Iran's digital economy are divided into eight branches: digital health, digital education, intelligent transportation systems, smart home, digital agriculture, digital tourism, fintech, and cyber security, showing digital agriculture with 117 products (8.6%) of all manufactured products is in the seventh place (**Figure 5a**) [23].

The statistics presented do not show the total products of the digital economy [23] but give an overview of the digital agriculture situation in Iran. In the world, digital agriculture is not as prominent as other sectors of the digital economy. The distribution of products in different agricultural sectors shows that most products are related to the marketplace (61.5%) and agricultural intelligence (21.4%), respectively, which includes 82.9% of the total products (**Figure 5b**). Agriculture intelligence includes smart agriculture, smart animal husbandry, smart poultry, smart farm, smart irrigation, and smart aquaculture. This statistic is not clear and transparent because greenhouse smartening is perhaps the most important part of smartening in Iranian agriculture, which in this statistic is probably a subset of smart agriculture. On the other hand, companies that produce greenhouse automation products are also active in smartening mushroom breeding halls, poultry, and livestock farms. Of course, as mentioned before, this statistic can show the ratio of different products in the digital agriculture sector. To study digital agriculture in Iran, we survey smart greenhouse, smart irrigation, drones in agriculture, and robotics.

3.1 Digitalization toward smart greenhouses

Iran's greenhouse area increased from 600 hectares in 1996 to 15,700 hectares in 2019. In the last decade, the average annual growth of the greenhouse cultivation area

Figure 6.
Automation and control of greenhouse using PLC.

in Iran has been about 1000 hectares. The development of greenhouses has made it attractive to invest in related industries, including greenhouse automation. In many cases, agricultural graduates have priority for the greenhouse construction, or an agricultural expert is required to work as a greenhouse consultant. Due to the employment of agricultural graduates, the demand and acceptance of new technologies in the Iranian greenhouse industry are easier than in other parts of it.

Studies have been conducted on the greenhouse automation system manufacturing and evaluation in Iran [30–32]. AS the evaluation of commercial greenhouse automation systems has not been carried out in Iran and, there are no data in this regard, to check the status of greenhouse automation systems, in addition to visiting some greenhouses with automation, interviewing was done with some manufacturers and greenhouse owners. In Iran, due to the existing market, several companies are currently making the automation systems of climate, feeding, irrigation, carbon dioxide injection, and lighting for greenhouse. The performance of automation systems can be evaluated from both hardware and software (control algorithm used in them). In terms of hardware design, manufactured systems are generally based on PLC (Programmable Logic Controller) (**Figure 6**), and manufacturers rarely design their specific electronic boards for greenhouse automation systems.

The reason is that the market is practically small because not all greenhouses request the installation of an automation system, and as long as companies are not sure that they have the right number of orders, the design and implementation of the board is not economically justified. One of the first companies that make its specific electronic boards is not able to send SMS (Short Message Service) to its clients with 3G or 4G of wireless mobile telecommunications due to the old hardware of the board and lack of updates. Also, the operation of the electronic boards is not stable and sometimes issues error commands. In many greenhouse climate control systems, the central controller communicates by the sensors and actuators via wires (**Figure 7**).

Figure 7.
Demonstration of sensor placement inside greenhouse environment.

The manufacturers believe the metal structure of the greenhouse blocks the wireless connection like a shield or reduces the antenna's field of view. Also, the height and the moisture content of the plant, as well as the ambient humidity, can damage wireless data transmission. Rezvani et al. [33] also pointed out that water in the high amount of biomass of the plants damps the radio signals and avoids communication distances over long ranges. Of course, the poor performance of wireless sensors in some projects has affected the mindset of greenhouse owners. Approximately one sensor is installed per 1000 square meters. But the number of temperature and humidity sensors is not equal, and the number of temperature sensors is almost twice as many as that of the humidity sensors. As a result, the relative humidity distribution or vapor deficit pressure (VPD) cannot be monitored like the temperature at the greenhouse surface. The ability to connect and control greenhouse equipment with the Internet (Internet of Things) and send messages via SMS to the operator is available in almost all greenhouses with automation. Of course, in some cases, by disconnecting the server of the support company, sending the text message to the mobile is practically stopped.

In large greenhouses (two or three hectares), the transmission of sensors data through wiring creates operational problems and increases the number of masters for data collection and processing. Using this method is very costly and time consuming, especially in places where the distance between sensors and actuators to the central board is long, and for this reason, researchers always try to reduce the consumption of wires and cables by using specific methods. One of the methods is to use a bus line so that all sensors and actuators are connected to the central board through a single cable [34]. The transmitter and receiver system or network connection are the most useful controlling method, especially effective for control operations that require data collection from different points in large areas [9, 17]. The main of remote technologies is the ability to be controlled by an intelligent remote-control system and Internet connection module. The remote-control systems have limitations in use and cannot be used easily and cheaply for all control purposes, especially the needs of the greenhouse. Therefore, researchers such as Jalilian et al. [35] use a wireless sensor network for designing greenhouse automation system.

Javadi Moghadam [31] successfully used the Zigbee transceiver to send data from temperature and humidity sensors in the greenhouse to Arduino boards for monitoring and IoT purposes. The climate control system was divided into three types of hardware including node, sensor, actuator, and central control or hub. The system consisted of two sensor nodes and, on the microcontroller board, a transmitter module was installed that was responsible for sending data to the central board. The sensor nodes used the Arduino Mega 2560 board, which contained a microcontroller with an AT Mega 2560 processor. An XBee S2 transceiver was used in each sensor so that it was possible to create cloud sensors. The temperature and humidity sensor used in each sensor node was DHT 22. The temperature and humidity data were called through a sensor connected to the board and sent to the central system via a radio transmitter (**Figure 8**). The range of the transceiver model is about 25 to 30 meters, which can be increased by about 10 meters by changing the UFL antenna to SMA.

One of the problems in greenhouse climate control systems is the lack of in-depth knowledge of greenhouse climate parameters and the interaction between the plant and the environment by the manufacturers of greenhouse automation systems. In almost all cases, the setting points include only temperature and relative humidity and no VPD control. The control algorithm is often on and off, and the PID (Proportional Integral Derivative) is not used. For this reason, sometimes available systems do not

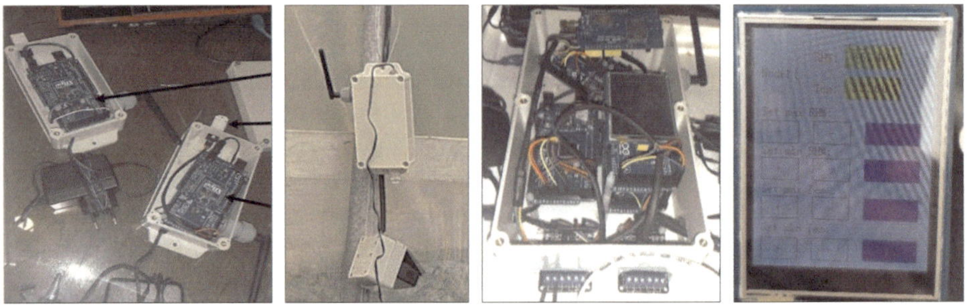

Figure 8.
An prototype automation system for small-scale greenhouses [31].

work well. Also, despite the use of the Internet of Things and metadata space, it is not provided to the user in an analyzed form. In limited cases, the algorithm for controlling the climatic parameters of the greenhouse has an error. Other problems include the lack of structures and suitable climate control equipment. If the ratio of ventilator opening to greenhouse floor area is not enough or climate control equipment such as heating and cooling systems are not appropriately designed and implemented or do not have the correct location, greenhouse automation systems will not work well.

3.2 Smart irrigation

Smart irrigation in agricultural fields is being developed in three approaches. In the first method, a platform is used for collecting data such as water right, soil properties, water source discharge, cultivation pattern, crop characteristics (length of cultivation period, crop coefficient), cultivation area (using satellite maps, Google Earth), irrigation system, irrigation frequency and costs, and revenues. Preparation and processing of meteorological information anywhere using interpolation from synoptic meteorological stations located in and around the zone, finally information analysis and estimation of required water and irrigation schedule offer and estimated yield to the farmer, are via SMS or website (**Figure 9**) [36]. In the second method, sensors of soil moisture, temperature and relative humidity of the environment, and wind speed are installed in the field (**Figure 10**). The amount of plant evapotranspiration is calculated by receiving environmental information by sensors and online data of the meteorological stations. The amount of crop water requirement is calculated based on the field climatic conditions, irrigation frequency, and type of cultivation. The farmer can irrigate his farm automatically or manually.

The system is based on IoT, and the user can log in to the system website online at any geographical point and while viewing a variety of graphic reports, he can be aware of the system's operational status and control irrigation remotely with his mobile phone. The sensors used in these systems are not wireless.

The third method is based on mobile application or device and like the first method can be used to calculate water needs with field data and meteorological data, but it is possible to install various sensors such as soil moisture or temperature and relative humidity of the environment. The system can perform the calculation based on the data collected from the sensors. In regions where the Internet is not available, the data are transferred to the mobile phone or device via Bluetooth, and after arriving the area where the Internet is available, and the data are analyzed and made available to the

Figure 9.
Screenshot of the homepage of the ibbrain.com, the first real smart irrigation for Iran [36].

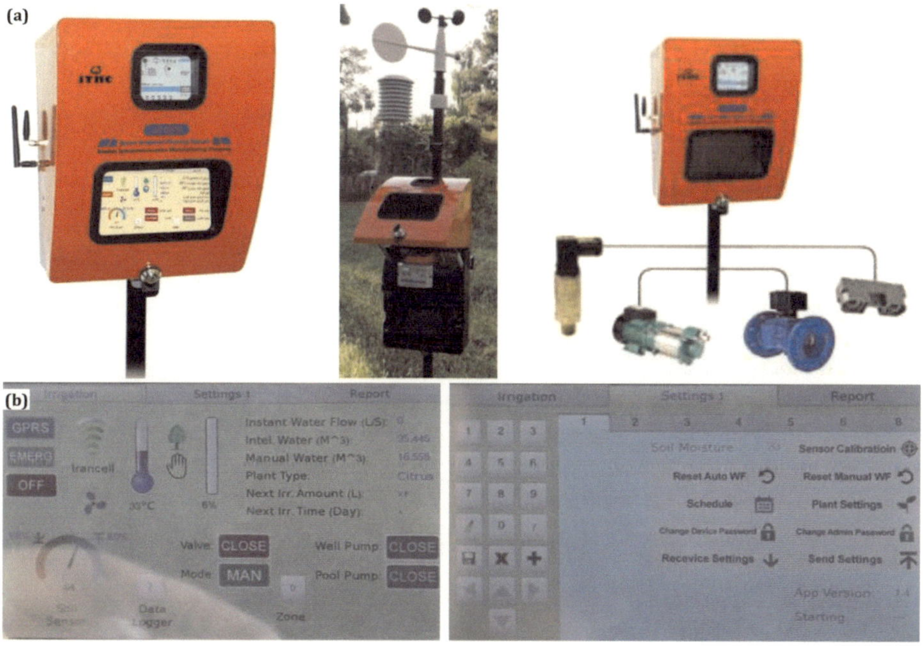

Figure 10.
Equipment (a) and monitoring (b) of the second approach smart irrigation.

user. The system also uses artificial intelligence and learning machines for better estimates.

Most farmers are not familiar with information knowledge, and the high cost of installing smart hardware-based irrigation systems on farms and their maintenance along with a small area of farmland and orchards and lack of full Internet coverage in rural areas make it difficult to develop smart irrigation systems. The mentioned factors have led to the development of platform-based approaches that do not require the installation of any hardware on farms and determine the volume of water and irrigation schedule from meteorological information and soil water balance. The costs of this method are much lower. In platform-based approaches, all the data are analyzed on the server and then provided to the farmer, and in case of interruption or failure of the server, the user's access to information is cut-off. Of course, there are backup servers, but due to exchange rate fluctuations, companies have problems renting servers or providing services.

3.3 Robotics

Although robots are not used in the agricultural sector of Iran, there is some research on the ir use in farming [37–39]. In Iran, a lot of research has been done on the development and efficiency of agricultural robots, especially in greenhouses. One of the fields of robotic research in greenhouses is a positioning system that can be classified as follows [38, 40]: Odometry; Inertial Navigation; Magnetic Compasses; Active Beacons; Global Positioning Systems; Landmark Navigation; and Model Matching. The positioning system was the most important research on agricultural robots, especially greenhouse robots, and is still one of the most important issues related to greenhouse robotics in Iran. Greenhouse sprayers are another research field on the usage of greenhouse robots [41]. Maneuvering and controlling these bots has created a fundamental challenge in greenhouse robots. Hence, researchers like [42] tried to solve this problem using mechanical manipulation robots.

Masoudi et al. [39] designed and constructed a three-wheel differential steering vehicle to act as the greenhouse sprayer (**Figure 11a**). Power was transmitted from two DC motors to two drive wheels through a gearbox and shaft system. A proportional controller was developed and tested to control the left and right motors, which navigated the aisles using the information provided by ultrasonic sensors. The robot was tested inside a greenhouse along a U-shaped path 0.98 m in width. Spraying, safety, and obstacle detection units of the vehicle were evaluated. The accuracy of the spray function was 99.47% and, the no-spray function was 99.92%, which is acceptable for greenhouse applications.

Haidari and Amiri Parian [38] designed and constructed a four-wheel differential steering mobile robot to act as a greenhouse robot (**Figure 11b**). The robot navigation was evaluated at different levels and actual greenhouses. The robot navigation algorithm was based on path learning so that the route was stored in the robot memory

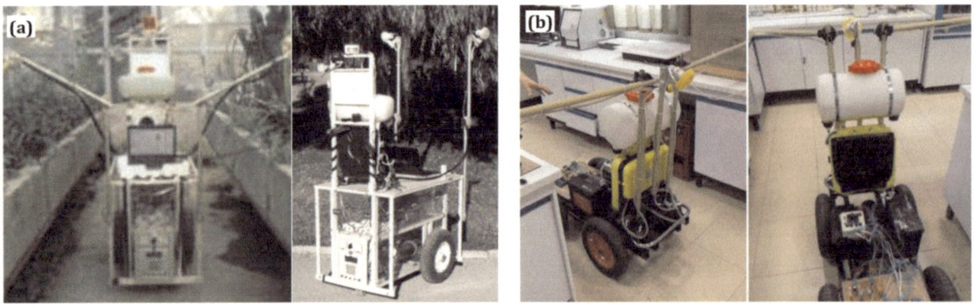

Figure 11.
Robots designed by Masoudi et al. [39](a) and Haidari and Parian [38] (b).

Figure 12.
Agricultural drone is spraying a farm [44].

using a remote control based on the pulses transmitted from the wheel encoders; then, the robot automatically traversed the path.

Gezavati, et al. [37] designed and built a precision seed planting robot for planting trays. First, based on the parameters designed in the laboratory, a prototype of wind seed planting was simulated by SolidWorks design software, and it was then constructed and evaluated for tomato seed planting. The planter consists of several parts operating harmoniously to yield the desired results. These parts include a chassis and conveyor belt mechanism, primary and secondary fertilizer tanks, squashing unit, seed metering device, and vibrating reservoir of the seed. The results showed that the nominal capacity of the seed robot was between 17,000 and 35,000 cells per hour. The accuracy of the designed system was 88% on average, and the nominal seed planting capacity of the system was 170 trays per hour. Drones have also been considered a specific field in agricultural robots. Shahrooz et al. [43] developed drone research to spray agricultural land in Iran. The production and sale of these robots require strong companies with appropriate services and support. Agricultural drones are often used for spraying and foliar spraying of farms (**Figure 12**). The most important problem of using UAVs (*Unmanned Aerial Vehicle*) is the price and depreciation of lithium polymer batteries. Security restrictions on obtaining flight permits are another problem with the use of drones in agriculture. The Ministry of Agriculture-Jihad supports the use of UAVs in the agriculture sector, and thus, UAV market is developing.

4. Challenges of digital agriculture in Iran

This section addresses the major challenges facing digital agriculture in Iran. The great majority of Iran's agriculture sector is in the agricultural 2.0 (combustion engine power) stage and requires extensive investment and training to transition to digital agriculture (agriculture 4.0). However, the study of Iran's budget bills indicates that despite the great emphasis on the importance of the ICT sector, from 2015 to 2018, the share of this sector to the total public budget declined, in a way that it reached from 3.6% to 2.4% and in the budget bill of 2019, it was similar to the previous year [45]. The Network Readiness Index (NRI) is another criterion for assessing the status of ICT use in countries. According to the global information technology report in 2016, Iran ranks 92nd among the 139 countries surveyed in this index and has acquired scores 3.7 out of 7 (the best status). Iran has the worst ranking in NRI in the pillar of the use of information and communication technology by companies (business usage), while one of the requirements for the realization of the digital economy is the increase in the use of digital technologies by businesses. Iran is in an unfavorable position compared with other countries, and its distance from the top countries in the MENAP region (Middle East, North Africa, and Pakistan) is very significant [46].

In Iran, 38 different documents related to information and communication technology and the digital economy have been compiled. Examination of these documents shows that the prevailing view of these documents is the field of ICT as public infrastructure, and less attention has been paid to it as a tool to create value in various industries and create new businesses that can hinder the development of the digital economy [45]. The similarities and overlaps of numerous and different institutions in the ICT and digital economy functions and tasks with parallelism, the overlap of activities, and lack of integration in policy making are other challenges in the development of the digital economy in the country. Other important issues are closing the legal gaps related to the ICT sector and adapting the laws and regulations of the country to the digital economy, especially in the discussion of privacy and information protection [45].

Small farmers suffer from lack of infrastructure and resources in rural areas and face challenges that limit their productivity and income. The low information knowledge of farmers is one of the most important reasons for preventing technology development in the agricultural sector. The smart and commercial systems on the market have complex instructions and farmers cannot get acquainted with how these systems work. Non-specialized policies in the development of smart agricultural, high initial cost and maintenance costs, and lack of appropriate support services have made smart systems less popular among subsistence farmers. The skilled and capable workforce is one of the main pillars of the formation of the digital economy so that the lack of human capital in Iran has become one of the obstacles to the creation and development of the digital economy.

5. Conclusions

The development of the digital economy is one of the most important development programs of the Iranian government. The digital economy share of GDP in Iran was 6.5% in 2019, and the goal is to reach 10% by 2025. Digital agriculture with 117 products (8.6%) of all manufactured products is in the seventh place of the digital economy. Most digital agriculture products are related to the marketplace (61.5%) and agricultural intelligence (21.4%), respectively, which include 82.9% of the total products. To study digital agriculture in Iran, we survey smart greenhouse, smart irrigation, and robotics. Approaches and their problems in greenhouses and smart irrigation were investigated. Studies on the use of robots in agriculture, often in the greenhouse sector, were also reviewed. Finally, the challenges facing digital agriculture such that most farmers are not familiar with information knowledge, the lack of necessary infrastructure in rural areas, the declining trend of investment in the budget sector in the food sector, and the need to reform laws and integrated management of the digital economy.

Author details

Seyed Moin-eddin Rezvani[1], Redmond R. Shamshiri[2], Jalal Javadi Moghaddam[3], Siva K. Balasundram[4] and Ibrahim A. Hameed[5]*

1 Agricultural Engineering Research Department, Hamedan Agricultural and Natural Resources Research and Education Center, AREEO, Hamedan, Iran

2 Leibniz Institute for Agricultural Engineering and Bioeconomy, Potsdam-Bornim, Germany

3 Department of Greenhouse Engineering, Agricultural Engineering Research Institute, Agricultural Research Education and Extension Organization (AREEO), Karaj, Iran

4 Department of Agriculture Technology, Universiti Putra Malaysia, Serdang Selangor, Malaysia

5 Faculty of Information Technology and Electrical Engineering, NTNU, Department of ICT and Natural Sciences, Ålesund, Norway

*Address all correspondence to: ibib@ntnu.no

References

[1] Bank TW. World and Iran indicator data. Available online: https://data.worldbank.org/indicator [Accessed: January 8, 2022]

[2] Agriculture, I.S.f.P. Precision Agriculture Definition. Available online: https://www.ispag.org/about/definition [Accessed: December 21, 2021]

[3] Sponchioni G, Vezzoni M, Bacchetti A, Pavesi M, Renga F. The 4.0 revolution in agriculture: A multi-perspective definition. In: Proceedings of Summer School F. Turco-Industrial Systems Engineering; Brescia, Italy; 2019. p. 143-149

[4] Trendov M, Varas S, Zeng M. Digital Technologies in Agriculture and Rural Areas. Rome: FAO; 2019. p. 152

[5] Shamshiri RR, Balasundram SK, Weltzien C. Use cases of digital agriculture. In: Proceedings of Global Forum for Innovations in Agriculture. Abu Dhabi, UAE: The International Center for Biosaline Agriculture (ICBA); 2019. p. 22

[6] Shamshiri RR, Weltzien C, Hameed IA, Yule I J, Grift T E, Balasundram SK, et al. Research and development in agricultural robotics: A perspective of digital farming. International Journal of Agricultural and Biological Engineering. 2018;**11**(4):1-14

[7] Shamshiri RR, Hameed IA, Pitonakova L, Weltzien C, Balasundram SK, Yule I J, et al. Simulation software and virtual environments for acceleration of agricultural robotics. Features Highlights and Performance Comparison. 2018;**11**(4):15-31

[8] Shamshiri RR, Kalantari F, Ting KC, Thorp KR, Hameed IA, Weltzien C, et al. Advances in greenhouse automation and controlled environment agriculture: A transition to plant factories and urban agriculture. International Journal of Agricultural and Biological Engineering. 2018;**11**(1):1-22

[9] Shamshiri RR, Bojic I, van Henten E, Balasundram SK, Dworak V, Sultan M, et al. Model-based evaluation of greenhouse microclimate using IoT-sensor data fusion for energy efficient crop production. Journal of Cleaner Production. 2020;**263**:121303

[10] Salam A. Internet of things in agricultural innovation and security. In: Internet of Things for Sustainable Community Development. Cham: Springer; 2020. p. 71-112

[11] Da Silveira F, Lermen FH, Amaral FG. An overview of agriculture 4.0 development: Systematic review of descriptions, technologies, barriers, advantages, and disadvantages. Computers and Electronics in Agriculture. 2021;**189**:1-20

[12] Anonymous I. E-Agriculture Strategy Guide: Piloted in Asia-Pacific Countries. Bangkok, Thailand: FAO and International Telecommunication Union (ITU); 2016. p. 205

[13] Bagheri N, Kafashian J. Precision Agriculture: Advantages, Challenges, Opportunities; 52584. Iran, Karaj: Agricultural Engineering Research Institute; 2018. p. 50

[14] Rezvani SM-ED, Shamshiri RR, Hameed IA, Abyane HZ, Godarzi M, Momeni D, et al. Greenhouse crop simulation models and microclimate control systems, a review. In: Next-Generation Greenhouses for Food

Security. London, UK: IntechOpen; 2021. p. 1-22

[15] Shamshiri RR, Hameed IA, Balasundram SK, Ahmad D, Weltzien C, Yamin M. Fundamental research on unmanned aerial vehicles to support precision agriculture in oil palm plantations. In: Zhou J, Zhang B, editors. Agricultural Robots-Fundamentals and Application. IntechOpen; 2018. p. 91-116. DOI: 10.5772/intechopen. 80936

[16] Shamshiri RR, Hameed IA, Karkee M, Weltzien C. Robotic harvesting of fruiting vegetables: A simulation approach in V-REP, ROS and MATLAB. Proceedings in Automation in Agriculture-Securing Food Supplies for Future Generations. 2018;**126**:81-105

[17] Shamshiri RR, Hameed IA, Thorp KR, Balasundram SK, Shafian S, Fatemieh M, et al. Greenhouse automation using wireless sensors and IoT instruments integrated with artificial intelligence. In: Shamshiri RR, editor. Next-generation greenhouses for food security. IntechOpen. 2021. p. 1-23. DOI: 10.5772/intechopen.97714

[18] Sultan M, Ashraf H, Miyazaki T, Shamshiri RR, Hameed IA. Temperature and humidity control for the next generation greenhouses: Overview of desiccant and evaporative cooling systems. In: Shamshiri RR, editor. Next-generation greenhouses for food security. IntechOpen. 2021. p. 1-23. DOI: 10.5772/intechopen.97273

[19] Li M, Shamshiri RR, Schirrmann M, Weltzien C, Shafian S, Laursen MS. UAV oblique imagery with an adaptive micro-terrain model for estimation of leaf area index and height of maize canopy from 3D point clouds. Remote Sensing. 2022; **14**(585):1-28

[20] Shamshiri RR, Weltzien C. Development and field evaluation of a multichannel LoRa sensor for IoT monitoring in berry orchards. In: Meyer-Aurich A, Gandorfer M, Hoffmann C, Weltzien C, Bellingrath-Kimura S, Floto H (Hrsg.) 41. GIL-Jahrestagung, Informations-und Kommunikationstechnologie in kritischen Zeiten. Bonn: Gesellschaft für Informatik e.V; 2021:289-294

[21] United Nations Conference on Trade Development. Digital Economy Report 2019: Value Creation and Capture: Implications for Developing Countries. NY: United Nations Publications New York; 2019

[22] Bukht R, Heeks R. Defining, Conceptualising and Measuring the Digital Economy. Development Informatics Working Paper 68. UK: University of Manchester; 2017. p. 1-24

[23] Mohammadi M, Shaifian A, Hashempoor A, Yeganeh Far A, Kakavandi Y, Lotfi I, et al. IRAN digital economy report 2020. Center for Monitoring and Promoting Digital Economy (Saytek); Iran; 2021

[24] Ministry of I.T.C (Iran). Statistics and information of digital economics. Available online: https://www.ict.gov.ir/fa/bishtar/measuring [Accessed: December 1, 2021)

[25] Ministry of I.T.C (Iran). Forty-eighth IoT expert critique and Opinion Meeting report. Available online: https://wa.ict.gov.ir/en/newsagency/24484/ [Accessed: December 25, 2021]

[26] Tanumand MR. Iran's IoT Infrastructure Development is not Astisfactory. Available online: https://www.ilna.news/fa/tiny/news-685708 [Accessed: October 30, 2021)

[27] Itbaz Editorial Team. IoT in Iran, The Study of the Internet of Things in Iran. Available online: https://itbaz.net/7966/iot-in-iran/ [Accessed: January 10, 2022]

[28] Orandi B. The Situation of the Internet of Things in Iran. Available online: http://www.majazi.ir/general_content/90521/swd_id/13825/unitWdId/53777/ [Accessed: January 7, 2020]

[29] Lotfi Z. IoT Challenges in Iran. Available online: https://bontech.ir/blog/ [Accessed: December 15, 2022]

[30] Zarei G, Azizi A. Design, construction and evaluation of a central device to control the environmental conditions of the greenhouse. Journal of soil and plant. Interactions. 2016;**6**:15

[31] Moghadam JJ. Design and Construction of a Greenhouse Environmental Control Device with the Ability to Control Hot Air Heaters and Ventilation. Karaj, Iran: Agricultural Engineering Research Institute; 2019. p. 43

[32] Omid M, Shafaei A. A computer-based control system for greenhouse environment: PART I-design and implementation of system. Journal of Agricultural Engineering Research. 2005;**6**:20

[33] Rezvani SM, Abyaneh HZ, Shamshiri RR, Balasundram SK, Dworak V, Goodarzi M, et al. IoT-based sensor data fusion for determining optimality degrees of microclimate parameters in commercial greenhouse production of tomato. Sensors. 2020;**20**:32

[34] Andres M, Nair R. A predictive-control framework to address bus bunching. Transportation Research Part B: Methodological. 2017;**104**:123-148

[35] Jalilian M, Sariri H, Parandin F, Karkhanehchi MM, Hookari M, Jirdehi MA, et al. Design and implementation of the monitoring and control systems for distribution transformer by using GSM network. International Journal of Electrical Power & Energy Systems. 2016;**74**:36-41

[36] Pirtaj HR. Agricultural Water Allocation Optimizer System: HOOSHAB System. Iran: Avand HooshAB; 2020. p. 1

[37] Gezavati J, Zamani DM, Abbasgolipour M, Mohammadi B, Randhi A. Preliminary design, construction and evaluation of robot of tomato seed planting for the trays of greenhouse. Journal of Agricultural Machinery. 2015;**5**:242-250

[38] Heidari A, Amiri Parian J. Greenhouse Mobile robot navigation using wheel revolution encoding and learning algorithm. Journal of Agricultural Machinery. 2021;**11**:1-15

[39] Masoudi H, Alimardani R, Omid M, Mohtasebi S, Bagheri S. Design, fabrication and evaluation of a mobile robot for spraying in greenhouses. Journal of Agricultural Engineering Research. 2011;**12**:87-100

[40] Borenstein J, Everett HR, Feng L, Wehe D. Mobile robot positioning: Sensors and techniques. Journal of Robotic Systems. 1997;**14**: 231-249

[41] Mosalanejad H. Evaluation of navigation system of a robot designed for greenhouse spraying. International Journal on Smart Sensing and Intelligent Systems. 2020;**13**(1):1-9

[42] Roshanianfard A, Noguchi N, Kamata T. Design and performance of a robotic arm for farm use. International Journal of Agricultural and Biological Engineering. 2019;**12**:146-158

[43] Shahrooz M, Talaeizadeh A, Alasty A. Agricultural spraying drones: Advantages and disadvantages. In: Proceedings of 2020 Virtual Symposium in Plant Omics Sciences (OMICAS). 23-27 Nov 2020; Bogotá, Colombia. New York: IEEE. p. 1-5. DOI: 10.1109/OMICAS52284.2020.9535527

[44] Behtuyi M. Available online: http://www.siranguav.ir [Accessed: February 6, 2022]

[45] Soltaninejad E. Economics in Simple Language: Digital Economy; 47. Iran: Iran Chamber Research Center; 2020. p. 30

[46] Baller S, Dutta S, Lanvin B. Global information technology report 2016. Ouranos Geneva: Innovating in the Digital Economy. World Economic Forum. 2016. p. 289

Chapter 5

Neutron-Gamma Analysis of Soil for Digital Agriculture

Galina Yakubova, Aleksandr Kavetskiy, Nikolay Sargsyan, Stephen A. Prior and Henry Allen Torbert

Abstract

This chapter describes technical aspects of neutron stimulated gamma ray analysis of soil carbon. The introduction covers general principles, different modifications of neutron gamma analysis, measurement system configurations, and advantages of this method for soil carbon analysis. Problems with neutron-gamma technology for soil carbon analysis and investigation methods including Monte-Carlo simulation of neutron interaction with soil elements are discussed. Based on investigation results, a method to extract the "soil carbon net peak" from raw acquired data was developed. A direct proportional dependency between the carbon net peak area and average carbon weight percent in the upper 10 cm soil layer for any carbon depth profile was demonstrated. Calibration of the measurement system using sand-carbon pits and field measurements of soil carbon are described. Compared to traditional chemical analysis (dry combustion) data, measurement results demonstrated good agreement between methods. Thus, neutron stimulated gamma ray analysis can be used for *in situ* determination of near surface soil carbon content and is applicable for precision geospatial mapping of soil carbon.

Keywords: neutron-gamma analysis, PFTNA, soil carbon storage, soil carbon distribution maps, scanning technology, ArcGIS, IGOR

1. Introduction

Agricultural operations play important roles in productivity and profitability of soil resources, and can influence aspects of climate change and the ability of soil to sequester carbon which has relevance to when quantifying carbon storage for credits. Development of sustainable land use practices requires understanding and evaluating impacts of these practices on soil resources. Exact knowledge of soil chemical composition can improve modern precision agricultural practices. For these reasons, large-scale measuring and mapping of soil elements (primarily carbon) on agricultural lands has become important today.

The current "gold standard" method of "dry combustion" for soil carbon determinations is based on extensive analysis of laboratory processed field samples. This method is labor intensive and time consuming. Other techniques (i.e., laser-induced breakdown spectroscopy, near- and mid-infrared spectroscopy, diffuse reflectance infrared Fourier transform spectroscopy, and pyrolysis molecular beam mass spectrometry) yield carbon values for small soil volumes (0.01–10 cm^3) near the soil surface [1], which may not be representative of large field areas.

Neutron-gamma analysis (NGA) can overcome these disadvantages and can be used to create soil carbon distribution maps of large field areas. Soil carbon content, expressed in average carbon weight percent in the upper 10 cm soil layer (*Cw%*), can be derived directly from *in-situ* neutron-gamma analysis results. Knowledge of soil density allows for calculating carbon content in mass units. The ratio of carbon in the 10–30 cm topsoil layer of various soil types can be used for estimating soil carbon to a 30 cm depth. This 0–30 topsoil layer is used for estimating carbon sequestered in soil [2].

Modified NGA, particularly Pulse Fast Thermal Neutron Analysis (PFTNA as classified by [3]), can be used for determining soil elemental (C, H, Si, etc.) contents. This can be accomplished by analyzing soil gamma spectra induced by 14.1 MeV neutron pulses. This includes gamma spectra acquired during neutron pulses (i.e., from gamma rays appearing due to inelastic neutron scattering, INS spectra) and between pulses (i.e., from gamma rays appearing due to thermal neutron capture, TNC spectra). Details concerning this methodology have been previously described [4–7].

A custom mobile PFTNA system was developed and constructed for measuring soil carbon in agricultural fields in the scanning regime [4, 6]. A GPS device and specially developed software were added to the mobile system for simultaneous acquisition of gamma signals and geographical positions. Maps of surface soil carbon distribution were constructed utilizing this system in conjunction with IGOR software [8] and ArcMap [9]. Technical aspects of neutron stimulated gamma ray analysis of soil carbon, developed algorithms, methodology and software for data acquisition, data processing, and mapping will be described in this chapter. In addition, factors that affect measurement error and required measurement times will be discussed.

2. Materials and methods

2.1 Physical basis of PFTNA

NGA is based on registration of gamma rays that appear in soil under neutron irradiation. A neutron generator is used as a neutron flux source. Each soil element issues gamma rays with predefined energies during certain nuclear reactions of that element with neutrons. Detectors register gamma rays as a spectrum that is the dependency of the registered gamma rays vs. their energy. In general, this gamma spectrum consists of many gamma peaks produced by various elements due to different processes of neutron-nuclear interactions and the continuous background. Since some peaks overlap, extraction of gamma peaks that correspond to particular soil elements is difficult. PFTNA can be used to overcome this problem. This method uses the difference in duration of INS (pico- and femto-second intervals) and TNC (dozens microsecond intervals) nuclear reactions to divide the spectra that appear due to these processes. With PFTNA, the neutron generator works in the pulse regime, and the single spectrum acquired is divided into two spectra in two separate memory locations. The INS spectrum, which is the gamma ray spectrum that appears due to inelastic neutron scattering of soil nuclei, is acquired during neutron pulses. The TNC spectrum, which appears due to thermal neutron capture, is acquired between pulses. Examples of these spectra are shown in **Figure 1**.

In the INS spectrum, the gamma peak used for determining soil carbon (centroid at 4.44 MeV) is still a complicated peak. This peak consists of the gamma response

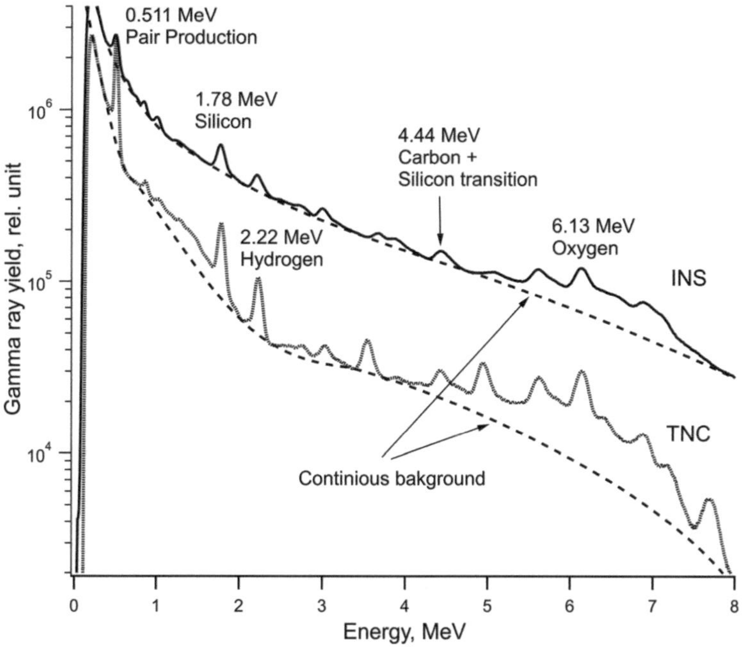

Figure 1.
Example of INS and TNC gamma spectra showing some peaks of interest.

from soil carbon, carbon in measurement system construction materials, and the cascade transition peak of silicon-28. A system background measurement should be conducted to define the gamma response corresponding to carbon in system components. This measurement is conducted under conditions where the effect of soil on the spectra is negligible (i.e., system is lifted to a height of more than 4–6 m above the ground). The silicon-28 cascade transition peak can be defined from determining values of the silicon-28 peak (centroid at 1.78 MeV in the spectra) and the cascade transition coefficient. The net carbon peak area can be computed by removing the background and silicon portions from the carbon peak. The net carbon peak area is directly associated with the average carbon content in the upper 10 cm soil layer expressed in weight percent [5]. This is true for any soil type regardless of carbon distribution shape. To relate net carbon peak area with average soil carbon content, corresponding calibration measurements (PFTNA measurements of model soil samples with well-known carbon contents) should be performed. Such measurements are needed to develop an equation for calculating soil carbon content from measured net carbon peak areas.

2.2 Carbon content returned by PFTNA

Since carbon distribution in soil is not uniform or known, several carbon content parameters can directly affect PFTNA measurements. Since soil carbon can sometimes be characterized as carbon surface density in the 30 cm soil layer, it was assumed that the PFTNA system acquired gammas from ~30 cm soil layer from irradiation by 14 MeV neutrons [1]. However, Monte-Carlo simulations did not confirm this assumption for unpredictable soil carbon distributions and soil densities [5].

Carbon content can be expressed as the average carbon weight percent in a given soil layer. Previous work showed [5] that soil carbon (expressed in weight percent in 10 cm upper soil layer) can be directly estimated from PFTNA gamma spectra measurements and corresponding peak area calculations. This can be done by applying previously defined calibration dependency using homogeneous reference samples. Workability of this expression for any type of soil with any soil carbon distribution shape with depth was confirmed using Monte-Carlo (Geant4, [10]) simulations. In addition, experimental measurements in agricultural fields were confirmed by comparison to traditional soil chemical analysis results.

2.3 PFTNA system design

To conduct soil carbon field measurements, a mobile PFTNA system was constructed on a platform (75 cm × 23 cm × 95 cm; ~300 kg) for towing by all-terrain vehicles over agricultural fields. The PFTNA system consisted of a MP320 pulsed neutron generator (NG; Thermo Fisher Scientific, Colorado Springs, CO), three 12.7 cm × 12.7 cm × 15.2 cm NaI(Tl) scintillation detectors (Scionix USA, Orlando, FL) with corresponding electronics (XIA LLC, Hayward, CA), a R2D-410 neutron detector (Bridgeport Instruments, LLC, Austin, TX), a power system (four 12 V 105 Ah DC105–12 batteries; a CGL 600 W-series DC-AC Inverter, Nova Electric, Bergenfield, NJ; and a PS4Quad Pro Charger, Pro Charging Systems, LLC, LaVergne, TN), a GPS device, an operational laptop, and an Android tablet. Iron and boric acid shielding is placed between the NG and gamma detectors to reduce irradiation of gamma detectors by fast neutrons (**Figure 2**).

The power system supplies all electronic equipment with 110 V AC voltage. Uninterrupted working time is ~20 h.

The neutron generator produces a pulsed output of 10^7–10^8 n s^{-1} depending on parameter settings; neutron energy is 14 MeV.

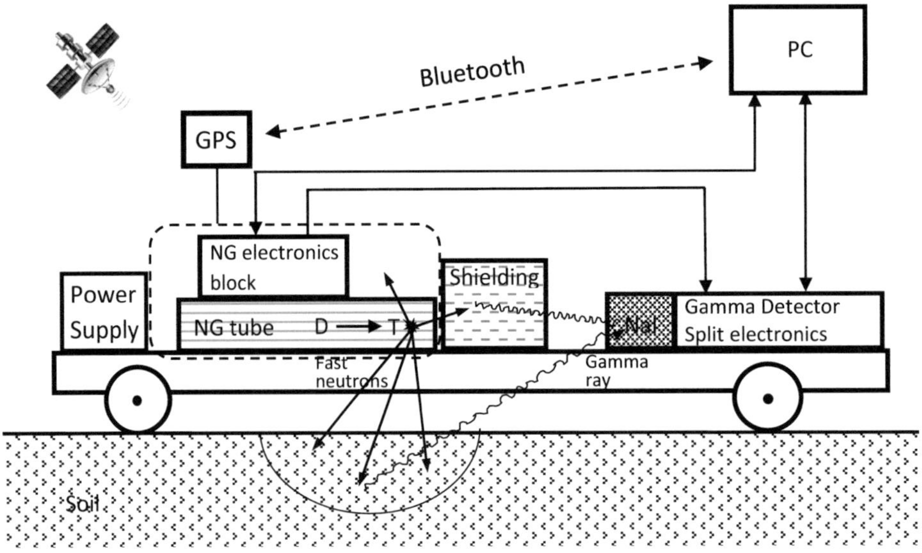

Figure 2.
Scheme of the PFTNA system.

Specially developed software allows spectral acquisition and defines the time interval for saving spectra to the laptop hard drive. This software also reads and saves GPS coordinates of the PFTNA system during scanning.

The tablet is mounted in the towing vehicle and is used for GPS set up and tracking system movement.

2.4 Data acquisition procedure

2.4.1 Gamma spectra acquisition

As mentioned above, the gamma peak with a centroid at 4.44 MeV is used to define soil carbon content. Gamma spectra containing peaks of interest with centroids at 4.44 and 1.78 MeV (used to correct silicon-28 interference in the carbon peak) are measured by gamma detectors. Gamma spectra measurements are the accumulation of gamma detector response in corresponding memory cells. Each memory cell accumulates the response (in counts) corresponding to a specific gamma ray energy interval. Gamma rays produced by 14 MeV neutrons have an energy less than this value. The studied energy interval is divided into 1024 cells (or channels) with channel widths of ~10 keV. Gamma ray production under neutron irradiation is a statistical process. Thus, spectra acquisition should continue for some time to achieve required accuracy. The average count rate (counts per second, cps) depends on neutron flux intensity, number of nuclei of interest in the sample (soil), efficiency of neutron-nuclei interactions, and detector(s) volume and efficiency of gamma ray registration. From a radiation safety viewpoint, total neutron yields exceeding 2×10^7 neutrons per second should not be used in a field system. In general, soil carbon content is no more than 5–10 w%, and agricultural soil density varies from ~1200 to 1600 kg/m^3. Under these conditions, gamma detectors with relatively large volumes should be used to achieve suitable count rates in channels of interest. In the described mobile PFTNA system, three gamma detectors with a total volume of ~7.5 dm^3 were used. To achieve a soil carbon content accuracy no worse than ±0.5 w%, the accuracy of carbon peak area determinations should be no worse than ±10 cps. The carbon peak area is around 200 cps when soil carbon content is ~2–3 w%. To reach the desired accuracy for the described equipment, measurement time should be no less than 15 min [11]. For elements having a soil content greater than carbon, determinations with this same accuracy require shorter spectra acquisition time. For example, the higher soil content of silicon (~30 w%) requires a spectra acquisition time of ~1 min or less.

2.4.2 Data acquisition modes

Soil carbon measurements using the PFTNA mobile system can be done in both static and scanning modes. In static mode, the system is moved to a particular position in the field, and measurements are performed for at least 15 min. Acquired data can be recorded at the end of measurement or periodically at desired time intervals. In scanning mode, the measurement system is continuously moved over the surveyed field, and acquired data are recorded every 30 s (or other previously defined time interval) during certain scanning time (see Section 2.4.4. for detail). Scanning mode is preferable for soil carbon measurement using the PFTNA mobile system since error associated with uneven soil carbon distribution at this scale (1–10 m) is practically negligible. Along with gamma spectra records, associated geographical coordinates defined by the GPS device are saved as well.

2.4.3 System background measurement and calibration

After construction, the PFTNA system should be calibrated prior to measuring soil elements. The calibration process consists of 2 parts: system background measurements and determining the dependency of the peak area of interest vs. elemental content in reference samples. This can be done for any soil element, but calibration for soil carbon content measurements are described herein.

System background is defined by peak areas of interest in the gamma spectra when the mobile system is lifted above the ground and away from any large objects. In this case, only system construction materials contribute to the gamma spectra. System background is part of the measured soil spectra and should be subtracted to attain the net soil spectra.

Reference samples for defining calibration dependency should be relatively large. For testing our PFTNA system, four 150 cm × 150 cm × 60 cm pits with sand-coconut shell mixtures of known carbon content (0, 2.5, 5 and 10 w% of carbon) were used. Calibration measurements should be performed such that errors are negligible compare to field measurements [4].

2.4.4 PFTNA field surveying methodology

To create soil elemental distribution maps, a number of evenly distributed points should be measured over the surveyed field. These can be represented in soil contour maps with elemental content isolines. Isolines can be created using Deterministic methods (Inverse Distance Weighting, Global polynomial interpolation, Local polynomial interpolation, Radial Basis Functions) or Geospatial methods (Kriging, Areal interpolation, Empirical Bayesian Kriging). Using these methods for surveying a field, there is a consensus that the required number of evenly distributed points (i.e., geographical coordinates and soil elemental content) needed for acceptable analysis is ~30, with 20 being the accepted minimum [12]. To attain this set of points, the surveyed field should be virtually divided into approximately equal site areas. Measurements can be done in static or scanning modes. If the field is believed to contain areas with sharp changes in soil elemental content (e.g., an asphalt road passing through the field), the number of sites (and therefore site area) should be adjusted accordingly.

To perform static mode measurements, the PFTNA system should be positioned at the center of each site for at least 15 min. In total, this mode would require a minimum of 5 h of measurement time excluding time required for moving the system between sites.

As previously mentioned, scanning mode measurements are preferable. In this mode, the system is towed within each site for ~15 min. The total measurement time is no different than static mode, but the error associated with unevenly distributed soil carbon is absent. To provide the required scanning time per site, the operator should select a suitable speed and path length. To aid the operator, the Android tablet installed in the cabin of towing vehicle traces the scanning path and displays the time spent at each site.

2.4.5 Soil density measurement

Results from PFTNA soil measurements is the average carbon weight percent in the upper 10 cm soil layer. To express soil carbon in mass units, soil density should also be concurrently measured to a depth of 10 cm (a Troxler 3440 Moisture Density Gauge aids in these measurements). Soil density is measured at five points in each site by the envelope scheme. The central point coincides with the geometric site center, and distance between points is ~40 m. Soil density for the site is assumed to be the average of the 5 points.

2.5 Data processing

2.5.1 Primary processing of spectra

The current FPTNA system has three gamma detectors. From a statistical point of view, processing each spectrum separately (peak areas calculation) and summarizing results of the three detectors is a common way of performing calculations. However, peak area determination from the spectrum of one detector yields relatively large statistical error since the soil carbon signal is relatively small. For this reason, spectra from the three detectors are summed prior to analysis.

During runtime, spectra acquired by each detector and corresponding geographic coordinates are saved at set time intervals. Each record (r) of raw data (for the ith detector, i = 1, 2, 3 detector number) consists of the following: measured INS and TNC gamma spectra $S_{INS,r,i}(Ch_{meas})$, $S_{TNC,r,i}(Ch_{meas})$, which are the number of counts in the channel (cnt/ch) versus channel number (Ch_{meas}) in the multichannel analyzer; real time of spectra acquisition ($RT_{INS,r,i}$, $RT_{TNC,r,i}$, s); input (absorbed by detector) and output (recorded in spectra) gamma ray count rates ($ICR_{INS,r,i}$, $ICR_{TNC.r,i}$, $OCR_{INS,r,i}$ and $OCR_{TNC.r,i}$, cps); clock time of recording of the INS and TNC spectra; and GPS coordinates. Due to each detector having its own energy calibration (correlation between photon energy and channel number), which can vary from day-to-day due to changing environmental conditions (primarily temperature), positions of peak centroids in spectra do not coincide (**Figure 3**).

Spectra of each detector must be brought to one energy calibration to be summarized. To achieve identical energy calibration, the energy calibration for a reference detector of the same type was established under laboratory conditions. To accomplish this by using several known gamma lines, the neutron stimulated gamma spectra (due to both inelastic neutron scattering and thermal neutron capture) of wet and dry soil, and soil-carbon mixes were acquired (see [4]). This resulted in several well-identified gamma peaks in the created spectra (e.g., 0.847 MeV iron peak, 1.779 MeV silicon peak, 2.223 MeV hydrogen peak, 4.438 MeV carbon peak, and 6.129 MeV oxygen peak, 7.63 MeV iron peak). These peak positions (in channel number) were used to create an energy calibration curve for the reference detector; this was a straight line in the range of interest. Spectra measured by other detectors (of the same type) under different conditions can be brought to this calibration line utilizing a shifting procedure (using Igor Pro software [8]).

With this procedure, channel numbers of two well identified peaks, $Ch_{1,meas}$ and $Ch_{2,meas}$, in each measured $S(Ch_{meas})$ spectrum are defined. Peaks with centroids at ε_1 = 1.78 MeV of ^{28}Si, and ε_2 = 6.13 MeV of ^{16}O (see **Figure 1**) are used. Next, channels of acquired spectra (Ch_{meas}) are shifted to a new position (Ch_{new}) (for all INS and TNC spectra) according to the following equations:

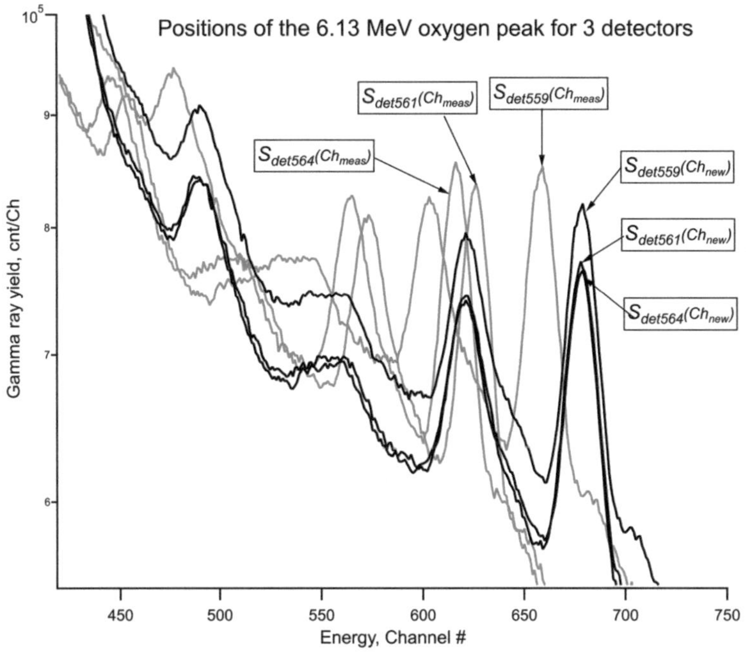

Figure 3.
Example of raw and shifted INS spectra of 3 detectors around the 6.13 MeV oxygen peak received during field scanning (559, 561, and 564 identify individual detectors in the PFTNA system).

$$Ch_{new} = \text{Int}[X(Ch_{meas})], \tag{1}$$

where

$$X(Ch_{meas}) = \frac{d_{ref} - d_{meas} + b_{ref} \cdot Ch_{meas}}{b_{meas}}, \tag{2}$$

$$d_{ref} = \varepsilon_1 - b_{ref} \cdot Ch_{1,ref}, \tag{3}$$

$$d_{meas} = \varepsilon_1 - b_{meas} \cdot Ch_{1,meas}, \tag{4}$$

$$b_{ref} = \frac{\varepsilon_2 - \varepsilon_1}{Ch_{2,ref} - Ch_{1,ref}}, \tag{5}$$

$$b_{meas} = \frac{\varepsilon_2 - \varepsilon_1}{Ch_{2,meas} - Ch_{1,meas}}, \tag{6}$$

$Ch_{1,ref}$ and $Ch_{2,ref}$ are the channel numbers for energy ε_1 and ε_2 in the reference calibration line. Count numbers in the channel with the new channel number $S(Ch_{new})$ are calculated as

$$\begin{aligned} S(Ch_{new}) = S(Ch_{meas}) - S(Ch_{meas}) \cdot \{X(Ch_{meas} - 1) - \text{Int}[X(Ch_{meas} - 1)]\} \\ + S(Ch_{meas} + 1)\{X(Ch_{meas}) - \text{Int}[X(Ch_{meas})]\}, \end{aligned} \tag{7}$$

Shifted spectra of the detectors (**Figure 3**) can be summarized. The shifted spectra are used in the next data processing steps.

2.5.2 Data processing static mode measurements

For static measurements, the PFTNA system is placed on a particular site where the carbon content must be defined. The required value for spectra acquisition time will depend on the desired statistical error. After spectra acquisition, the gamma spectra are shifted according to procedures described in Section 2.5.1. The net INS spectrum is found as the difference of summarized INS spectra (3 detectors) and summarized TNC spectra (3 detectors). The net INS spectrum (**Figure 4a**) is used for determining silicon (1.78 MeV) and carbon (4.44 MeV) peak areas. Peak areas are calculated by their Gaussian fitting using IGOR software [8]. The 1.78 MeV peak is approximated by one Gaussian (**Figure 4b**), while the 4.44 MeV peak uses two Gaussians (**Figure 4c**) since it contains a silicon transition component.

Received values of silicon ($PA1.78_{soil}$) and carbon ($PA4.44_{soil}$) peaks areas are used in the next steps of data processing for calculating of soil carbon content.

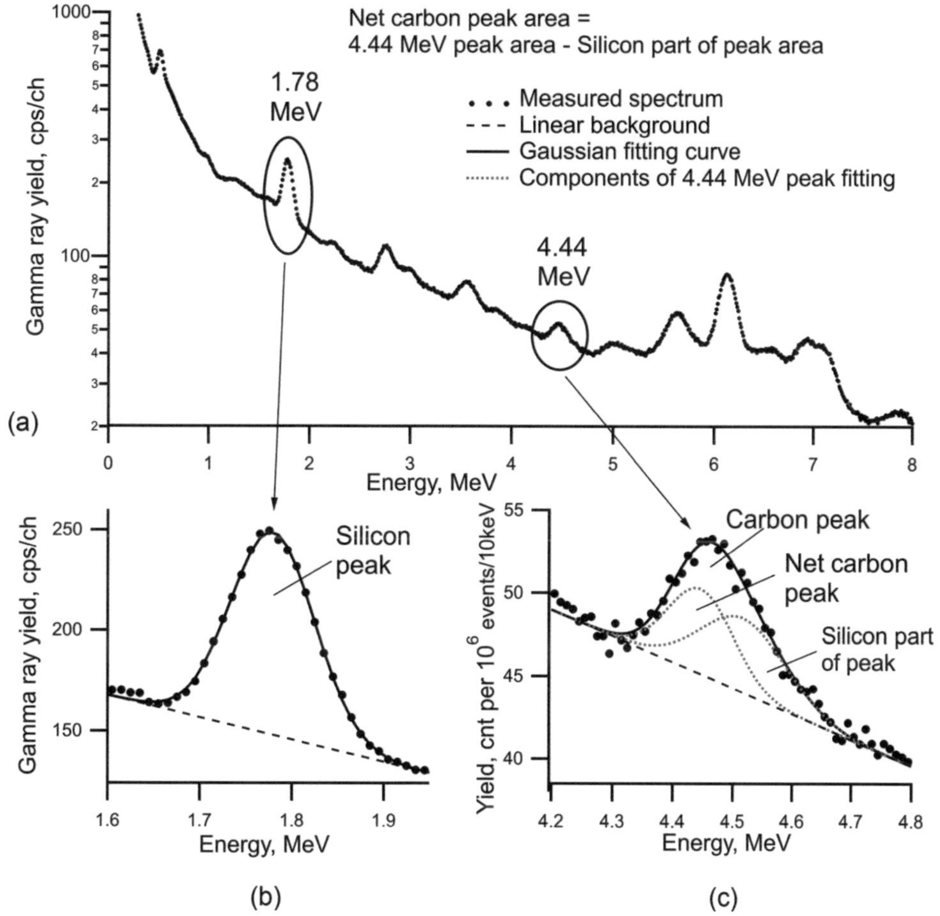

Figure 4.
Example of the net INS spectrum (a), and 1.78 MeV and 4.44 MeV peak fittings by one (b) and two Gaussians (c), respectively.

2.5.3 Data processing scanning mode measurements

When surveying in scanning mode, the PFTNA system is towed across the field while simultaneously measuring the gamma spectra. Acquired gamma spectra and geographical coordinates of the PFTNA system position are saved every 30 s (~50 m of travel). To ensure even coverage, the surveyed field is virtually divided into sites of approximately equal area. During scanning, the system should be present within each site for at least 15 min; this is required time ensures that error from the combined soil carbon spectrum attributed to each site (see further) not exceed 0.5 w% as explained in Section 2.4.2.

As previously mentioned, creating a map of soil carbon distribution requires a dataset consisting of no less than 20 points of defined elemental contents and corresponding geographical coordinates. To attain this dataset, the field should be virtually divided into the same number of sites. During data processing, the difference between two sequentially recorded spectra and geographical coordinate midpoints are determined, and the differential spectra (midpoints spectra) are assigned to these midpoints. All midpoint spectra having coordinates within a given site will be attributed to this site and after primary processing (as described in Section 2.5.1) will be averaged. The soil carbon content will be determined from this averaged spectrum. The dataset consisting of soil carbon content values and geographical coordinates of corresponding site centers will be used for creating maps.

All acquired spectra are processed on the data processing computer as follows. After spectra shifting procedures, gamma peaks at positions of interest become coincident in each spectrum. The differential spectra between two shifted sequentially recorded spectra for the ith detector, $\Delta S_{INS,r,i}(Ch_{new})$, $\Delta S_{TNC,r,i}(Ch_{new})$, are calculated (channel by channel) as:

$$\begin{aligned} \Delta S_{INS,r,i}(Ch_{new}) &= S_{INS,r+1,i}(Ch_{new}) - S_{INS,r,i}(Ch_{new}) \\ \Delta S_{TNC,r,i}(Ch_{new}) &= S_{TNC,r+1,i}(Ch_{new}) - S_{TNC,r,i}(Ch_{new}), \end{aligned} \tag{8}$$

where $S_{INS,r+1,i}(Ch_{new})$, $S_{TNC,r+1,i}(Ch_{new})$ and $S_{INS,r,i}(Ch_{new})$, $S_{TNC,r,i}(Ch_{new})$ are the shifted measured gamma spectra for $r + 1$th and rth record (in counts per channel) for ith detector and INS and TNC spectra, respectively. (Here and hereafter, all actions with spectra are done channel by channel).

The differential spectra in cps/ch (counts per second per channel), $\Delta S'_{INS,r,i}(Ch_{new})$ and $\Delta S'_{TNC,r,i}(Ch_{new})$ are calculated as:

$$\begin{aligned} \Delta S'_{INS,r,i}(Ch_{new}) &= \frac{\Delta S_{INS,r,i}(Ch_{new})}{LT_{INS,r+1,i} - LT_{INS,r,i}}, \\ \Delta S'_{TNC,r,i}(Ch_{new}) &= \frac{\Delta S_{TNC,r,i}(Ch_{new})}{LT_{TNC,r+1,i} - LT_{r,i}}, \end{aligned} \tag{9}$$

where $LT_{INS,r+1,i}$, $LT_{TNC,r+1,i}$ and $LT_{INS,r,i}$, $LT_{TNC,r,i}$ are the live time (in s) for the $r + 1$th and rth record for the ith detector, and INS and TNC spectra, respectively. Live time for each spectrum is calculated as [4]:

$$\begin{aligned} LT_{INS,r,i} &= RT_{INS,r,i} \cdot \frac{OCR_{INS,r,i}}{ICR_{INS,r,i}}, \\ LT_{TNC,r,i} &= RT_{TNC,r,i} \cdot \frac{OCR_{TNC,r,i}}{ICR_{TNC,r,i}} \end{aligned} \tag{10}$$

The two sums of the three differential spectra for each *r*th record, $\Delta S'_{INS,r}(Ch_{new})$ and $\Delta S'_{TNC,r}(Ch_{new})$, were then calculated as:

$$\Delta S'_{INS,r}(Ch_{new}) = \sum_{i=1}^{3} \Delta S'_{INS,r,i}(Ch_{new}),$$
$$\Delta S'_{TNC,r}(Ch_{new}) = \sum_{i=1}^{3} \Delta S'_{TNC,r,i}(Ch_{new}) \quad (11)$$

The net INS spectrum for each *r*th record $\Delta S'_{Net,INS,r}(Ch_{new})$ was then calculated as the difference between INS and TNC spectra as:

$$\Delta S'_{Net,INS,r}(Ch_{new}) = \Delta S'_{INS,r}(Ch_{new}) - \Delta S'_{TNC,r}(Ch_{new}) \quad (12)$$

The net INS spectra found in this manner will have geographical coordinates of corresponding midpoints. After sorting by site, the average spectra of all net INS midpoint spectra attributed to each site are found. Finally, these average spectra are used for determining soil carbon content for each site. This dataset consisting of soil carbon content and geographical coordinates of corresponding site centers will be used for creating maps.

2.5.4 Calculating soil carbon content

After primary processing of spectra (Section 2.5.1) and finding the summarized (3 detectors) INS and TNC spectra in counts rate (cps) and net INS spectra (Section 2.5.3), the peak areas of silicon (centroid at 1.78 MeV) and carbon (centroid at 4.44 MeV) can be found using Gaussian fitting procedures (in cps; Section 2.5.2). The background portions of these peaks were found as described in Section 2.4.3.

Carbon content (Cw%) is calculated by Eq. (13):

$$Cont_{\text{soil}} = \frac{\left(\text{PA4.44}_{soil} - \text{PA4.44}_{bkg}\right) - k_1 \cdot \left(\text{PA1.78}_{soil} - \text{PA1.78}_{bkg}\right)}{k_2}, \quad (13)$$

where PA4.44_{soil}, PA1.78_{soil} and PA4.44_{bkg}, PA1.78_{bkg} are the carbon and silicon peak areas in the soil and system background spectra, respectively, while k_1 is a silicon transition coefficient and k_2 is the calibration coefficient. These coefficients are defined during system calibration (see Section 2.4.3).

The total carbon content in the upper 10 or 30 cm soil layer of a surveyed field can be defined from PFTNA measurement results. In addition to PFTNA carbon content (in w%) data, field soil density (d in kg/m^3) is required. Determination of field soil density was described in Section 2.4.5.

Total field soil carbon in the 10 cm layer ($TC10$, ton) can be determined according to following equation:

$$TC10 = \sum_{i=1}^{n} \frac{Cont_{\text{soil } i}}{100} \cdot d_i \cdot S_i \cdot \frac{0.1}{1000}, \quad (14)$$

where n is the number of sites in a divided field for PFTNA measurements, $Cont_{\text{soil } i}$, and d_i, $S_i$are soil carbon content (w%), soil density (kg/m^3), and area (m^2) of the ith site, respectively. Area can be taken from the computer software used to divide the field into sites. Given that the PFTNA measurement result is an average soil carbon content for the field, $\overline{Cont_{\text{soil}}}$, then.

$$TC10 = \frac{\overline{Cont_{\text{soil}}}}{100} \cdot \overline{d_{\text{field}}} \cdot S_{field} \cdot \frac{0.1}{1000}, \tag{15}$$

where $\overline{d_{\text{field}}}$, S_{field} are average field soil density (kg/m^3) and field area (m^2), respectively.

Total carbon content in the upper 30-cm soil layer of the surveyed field ($TC30$, ton) can be defined as:

$$TC30 = \frac{TC10}{0.55}, \tag{16}$$

where the coefficient 0.55 is the ratio of the carbon surface density (g/cm^2) in the 10-cm layer to the carbon surface density in the 30-cm layer with an error of ± 0.10. This coefficient was found to be the average value for different carbon depth profiles for several agricultural fields in Alabama.

2.6 Measurement and data processing software

The system is supported by three software applications: Scanning App, Navigator App, and Computing App. The data flow within software applications is presented in **Figure 5**.

2.6.1 Scanning App

The mobile system is managed by the Scanning App. This Windows desktop application was developed in-house using the C# programming language and .Net WPF (Windows Presentation Foundation) technology [13]; this app can run on a consumer-grade computer. The Scanning App runs on the mobile system laptop; application features are presented in **Table 1**.

2.6.2 Navigator App

The map managing process is mainly performed through the Navigator App. The Navigator App is an Android application developed in-house with Kotlin programming language [16] and can run on a consumer-grade Android tablet or smartphone. Navigator App features are presented in **Table 2**.

2.6.3 Computing App

After measurement, the spectra from the Scanning App and the field boundary file from the Navigator App must be processed by the Computing App. The Computing App is a Windows desktop application developed in-house using the C# programming language and .Net WPF technology [13]; this app can run on a consumer-grade

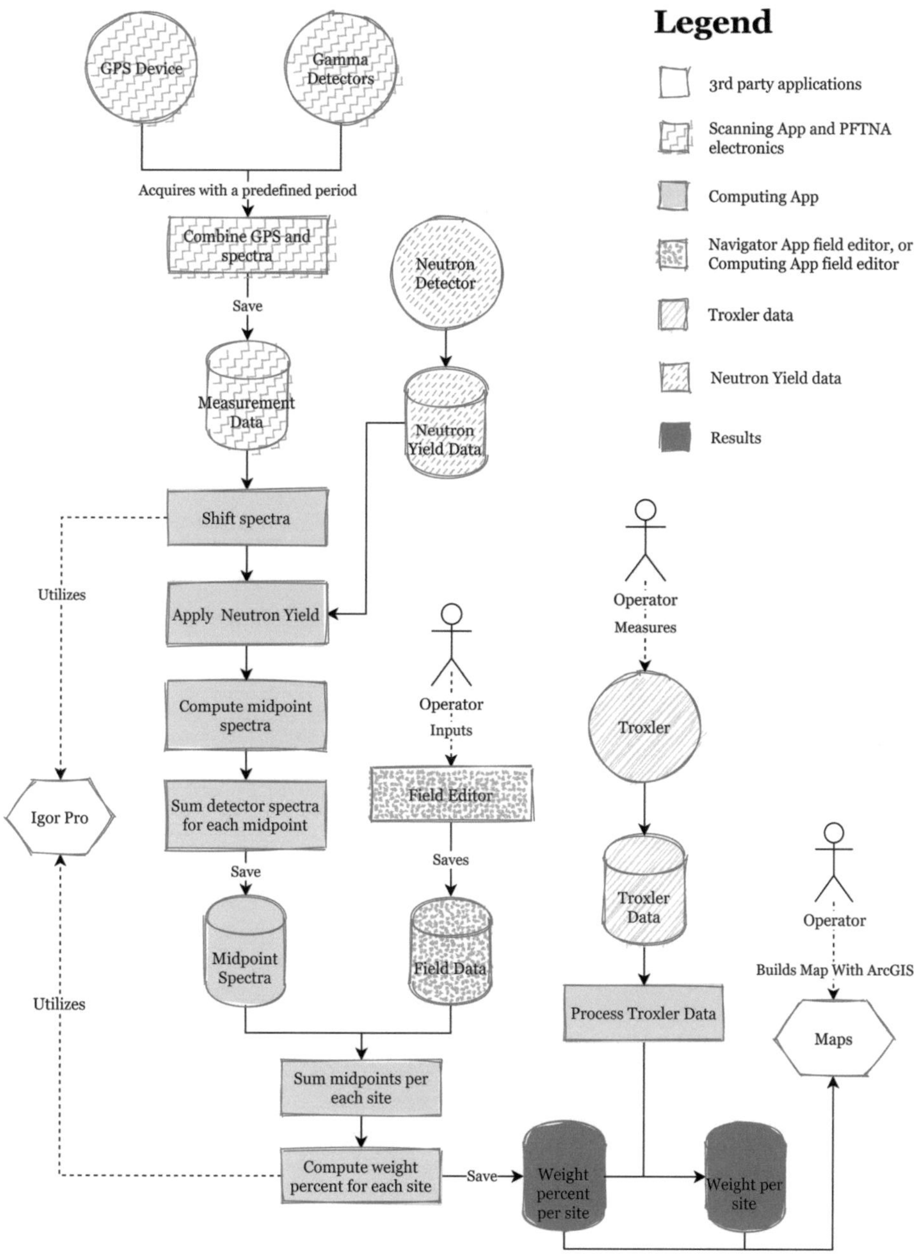

Figure 5.
Data flow within software applications.

computer. The Computing App implements the algorithms in Section 2.5.1–2.5.4 to process static and scanning mode spectra and produce carbon content results. For some mathematical operations on spectra, the Computing App automatically communicates with Igor Pro, which is a scientific data analysis software by WaveMetrics [8]. Additionally, the Computing App contains features presented in **Table 3**.

Feature name	Feature description
1. Gamma detector control	The current version of the Scanning App supports communication with system electronics. This app is not only capable of acquiring current spectra data from the gamma detector, but also provides an interface to access and edit all electronics settings required to tune the spectra acquisition process
2. GPS connectivity	The current version of the Scanning App supports any GPS device that can communicate with NMEA 0183 (National Marine Electronics Association) standard GLL (Geographic Position—Latitude/Longitude) or GGA (Global Positioning System Fix Data) protocols over a Bluetooth or USB port [14, 15]. The Scanning App can scan and automatically find the GPS device. GPS data is acquired in one second intervals
3. Spectra plot	The Scanning App features a plot that allows spectral zooming (in and out), adding guidelines, and loading spectra from saved files to allow the operator to visually analyze spectra
4. Adjustable time intervals	The time interval between spectra acquisitions can be customized. The time interval can also be set to increase logarithmically
5. Failure handling	The Scanning App pauses the measurement, alerts the operator via a detailed error message, and sounds an alarm in the event of the following scenarios: a. Neutron generator failure was inferred from the nature of the acquired spectra. The current version of the Scanning App cannot manage the neutron generator directly. b. Connection with the GPS device was lost. In this case, the Scanning App also constantly attempts to re-establish the connection. c. Connection with any of the detectors was lost. Due to the nature of the current detectors, the Scanning App must be terminated and manually restarted.

Table 1.
Scanning App features.

Feature name	Feature description
1. Creating field maps	The Navigator App allows the operator to create field maps consisting of multiple individual pieces (zones). For each zone, the number of sites in that zone can be defined, and the Navigator App will automatically generate the site polygons. For visual purposes, the operator can also adjust the color of the polygon boundaries
2. Editing field maps	The Navigator App allows editing of existing field map zone boundaries, adding or deleting zones, changing the number of sites in each zone, and color preferences
3. Scanning navigation	During the field scan, the Navigator App tracks the path of the mobile system, the time spent scanning, and the time spent at each individual site. Sites are also color-coded via a red-green-blue gradient scheme, with blue indicating that the site was scanned for the required time. Required time for sites can be set before scanning begins
4. Exporting and importing field maps	The Navigator App can export any field map or scanning map into a KML file. It also exports field boundary maps into a special file that can be imported into the Computing software or another Navigator App

Table 2.
Navigator App features.

Feature name	Feature description
1. Map management	The Computing App allows for modifying and exporting maps imported from the Navigator App
2. Troxler data support	Troxler Data can be imported and will be automatically distributed by sites and applied during computations. The Computing App outputs the weight of carbon (metric tons) for the upper 10 or 30 cm of soil
3. Additional data support	Additional data consisting of geolocation-value pairs can be imported and automatically distributed by sites
4. Neutron yield support	Neutron yield data can be imported for spectra correction
5. Additional analysis support	Apart from results data, the Computing App exports data corresponding to intermediary steps of the computing process, and outputs specially computed and formatted additional spectra data for further spectra analysis

Table 3.
Computing App features.

3. Results and discussion

Example soil carbon measurements conducted using the technology described in this chapter are presented in **Figures 6** and 7 and **Table 4**. These measurements were performed on a field in Iowa using the PFTNA mobile system. The field size was 53 ha. Scanning time was ~5.5 h.

Sites of equal area and the PFTNA scanning path are shown in **Figure 6**. Soil density measurement points and site centers are also shown. Geographical coordinates of site centers, values of carbon gamma peak areas, calculated values of soil carbon weight percent, and soil carbon content in the 10 and 30 cm layers for each site are presented in **Table 4**. The total carbon weight in the 10 and 30 cm layers of this field and the average carbon weight per ha are also shown in this table. The average carbon weight percent for this field was 3.45 w% with a variation (STD) of 0.44 w%. This variation is larger than the average error of soil carbon weight percent in each site, indicating that changes of carbon weight percent are present within the field. The carbon distribution map for this field was created using Local Polynomial Interpolation (Deterministic methods) in ArcMap based on Cw% site data (**Figure 7**). The insignificant change in carbon content from ~4 (east border) to ~3 w% (west border) can be seen on this field map. Knowledge of average values and carbon content changes across a field can be very useful in modern agricultural practices. Data regarding total carbon content in the 10 and 30 cm layers of this field can be useful for agricultural practice and ecological assessments.

Based on the discussed example and previous experiments, the equipment for implementing Pulsed Fast/Thermal Neutron Analysis of soil carbon content under field conditions was demonstrated to be reliable. Such measurements return soil carbon contents within a relatively short time for large fields (53 ha for ~5.5 h), and accuracy of measurements were no worse than traditional chemical analysis.

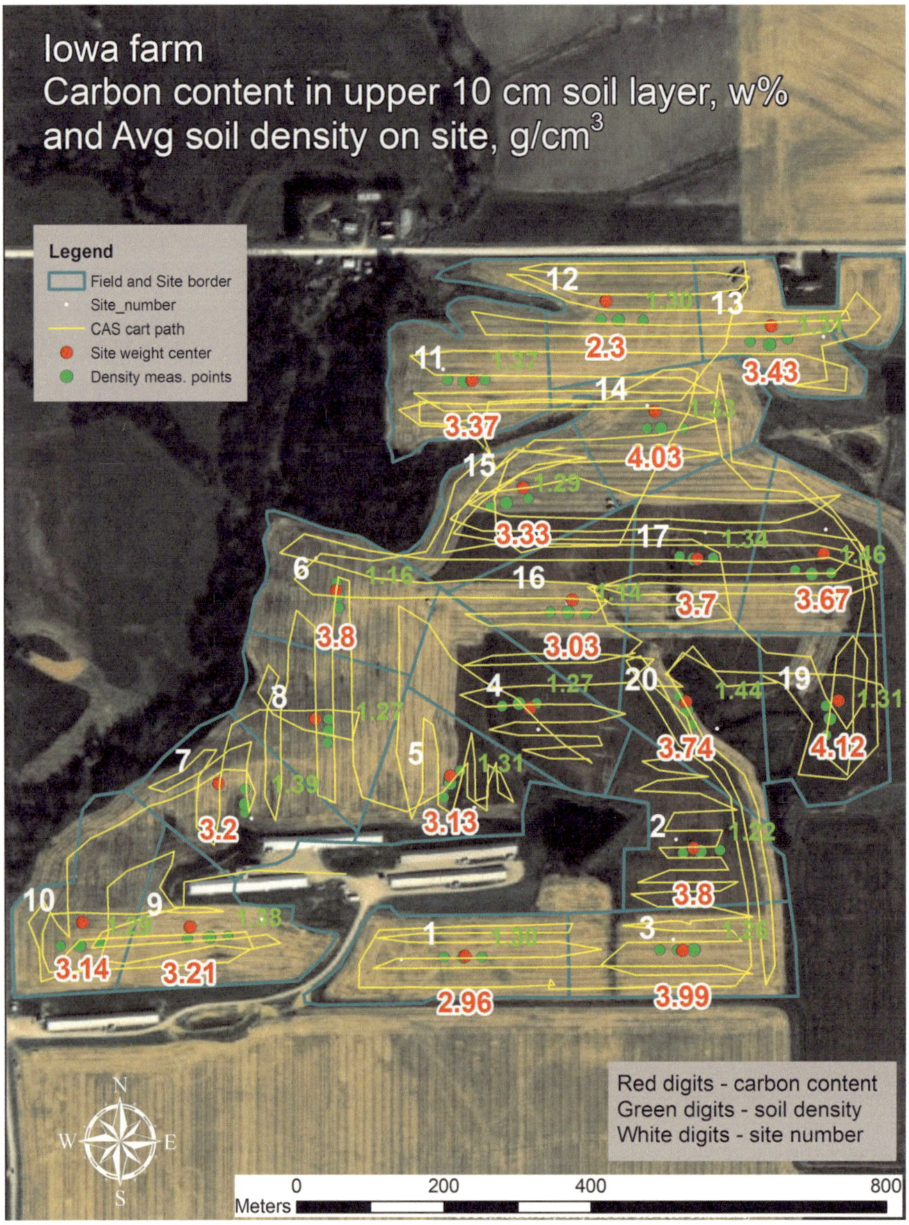

Figure 6.
Map showing field and site borders, scanning path, carbon content values, and site soil densities.

4. Conclusion

Application of neutron gamma analysis for soil elemental determinations can be an alternative to traditional chemical analysis. This technology has advantages over other methods since it is a nondestructive *in-situ* method that requires no soil sampling and associated laboratory processing.

The presented PFTNA methodology can be used for determination and mapping of soil carbon content. The accuracy of soil carbon analysis by PFTNA is no worse than

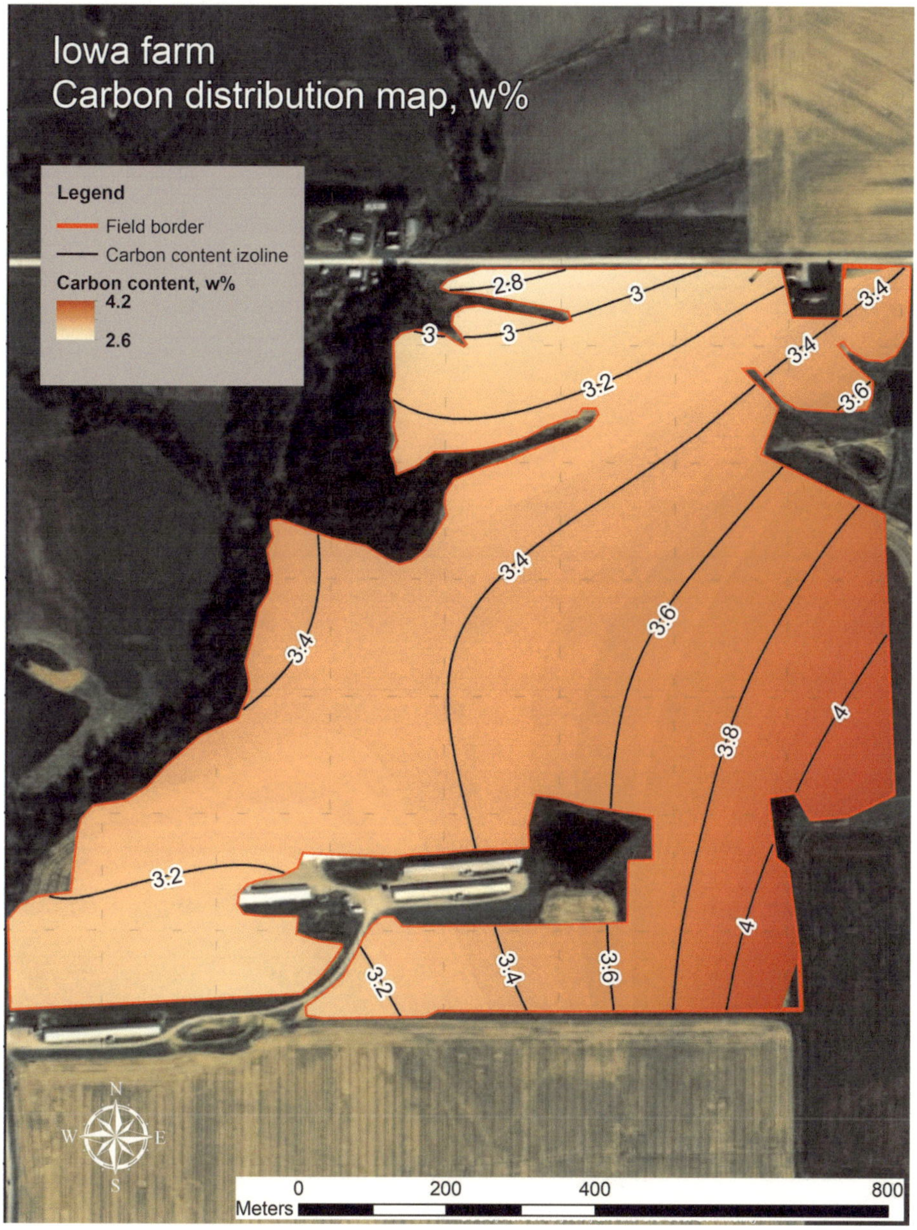

Figure 7.
Carbon distribution map.

traditional chemical analysis. Acquiring more experience and refining the described technology for large-scale soil carbon content determination under diverse field environments is the future direction of this research.

The equipment and methodology described in this chapter can also be applied to measure field content of elements such as Fe, Si, Al, H (water content) and Cl (soil contamination by chlorinated compounds). In addition, this mobile system can be used for measuring and mapping natural soil radioactivity, particularly potassium-40;

Site #	Latitude Longitude	Carbon peak area ± err, cps	Avg. carbon content in 10 cm ± err, w%	# of mid-points	Scanning time, m:s	Site area, ha	Avg. soil density in 10 cm layer, g/cm^3	Carbon content in 10 cm ± err, ton	Carbon content in 30 cm ± err, ton
1	41.2728 −92.0303	200 ± 6	2.96 ± 0.33	28	14:01	2.67	1.30	103 ± 12	187 ± 41
2	41.2731 −92.0275	218 ± 5	3.80 ± 0.28	38	19:00	2.61	1.22	121 ± 12	220 ± 46
3	41.2718 −92.0276	220 ± 5	3.99 ± 0.28	35	17:30	2.73	1.26	137 ± 10	249 ± 49
4	41.2747 −92.0295	217 ± 7	3.77 ± 0.37	31	15:30	2.91	1.27	139 ± 17	253 ± 56
5	41.2739 −92.0304	205 ± 7	3.13 ± 0.36	32	16:01	2.43	1.31	99 ± 12	181 ± 39
6	41.2761 −92.0318	216 ± 7	3.80 ± 0.35	31	15:32	2.67	1.16	118 ± 15	215 ± 47
7	41.2738 −92.0333	208 ± 6	3.20 ± 0.32	31	15:30	2.67	1.39	119 ± 21	216 ± 54
8	41.2746 −92.0321	208 ± 8	3.27 ± 0.41	30	15:01	2.67	1.27	111 ± 15	201 ± 46
9	41.2721 −92.0336	206 ± 6	3.21 ± 0.31	35	17:31	2.67	1.38	118 ± 15	214 ± 47
10	41.2722 −92.0350	203 ± 7	3.14 ± 0.37	31	15:30	2.67	1.29	108 ± 14	196 ± 44
11	41.2786 −92.0302	209 ± 6	3.37 ± 0.32	32	16:01	2.67	1.37	123 ± 13	223 ± 47
12	41.2796 −92.0285	187 ± 8	2.30 ± 0.41	31	15:30	2.75	1.30	82 ± 15	150 ± 38
13	41.2793 −92.0265	209 ± 9	3.43 ± 0.46	32	16:00	2.75	1.31	124 ± 18	226 ± 52
14	41.2783 −92.0279	220 ± 8	4.03 ± 0.44	31	15:31	2.58	1.33	138 ± 16	251 ± 54
15	41.2774 −92.0296	206 ± 6	3.33 ± 0.29	30	15:00	2.58	1.29	111 ± 12	202 ± 43
16	41.2760 −92.0290	201 ± 6	3.03 ± 0.33	32	16:00	2.67	1.14	93 ± 16	168 ± 42
17	41.2765 −92.0274	215 ± 7	3.70 ± 0.36	32	16:01	2.67	1.34	132 ± 16	240 ± 52
18	41.2766 −92.0259	213 ± 7	3.67 ± 0.38	28	14:00	2.67	1.46	143 ± 16	260 ± 55
19	41.2748 −92.0257	221 ± 6	4.12 ± 0.32	33	16:31	2.67	1.31	144 ± 15	261 ± 55
20	41.2748 −92.0276	216 ± 5	3.74 ± 0.29	34	17:01	2.67	1.44	143 ± 12	260 ± 52
Avg ± STD			3.45 ± 0.44						
			In 10 cm layer:			In 30 cm layer:			

Site #	Latitude Longitude	Carbon peak area ± err, cps	Avg. carbon content in 10 cm ± err, w%	# of mid-points	Scanning time, m:s	Site area, ha	Avg. soil density in 10 cm layer, g/cm^3	Carbon content in 10 cm ± err, ton	Carbon content in 30 cm ± err, ton
Total field carbon content ± error, ton			**2406 ± 66**			**4374 ± 216**			
Specific field carbon content ± error, ton/ha			**45 ± 1**			**82 ± 4**			

Table 4.
Results of calculating the carbon content of an Iowa field (confidence level of errors is 0.68).

in this case, the neutron generator is turned off since only gamma detectors are required. The application of the PFTNA technology for such assessments are other future topics of investigation.

Acknowledgements

This research was supported by U.S. Department of Agriculture-Agricultural Research Service National Soil Dynamics Laboratory and the authors are indebted to Mr. Barry G. Dorman and Mr. Robert A. Icenogle for technical assistance in experimental measurements, and to Mr. Dexter LaGrand for assistance with software installation. We thank XIA LLC for their electronics and detectors in this project.

Author details

Galina Yakubova[1], Aleksandr Kavetskiy[1], Nikolay Sargsyan[2], Stephen A. Prior[1] and Henry Allen Torbert[1*]

1 USDA/ARS/National Soil Dynamics Laboratory, Auburn, AL, USA

2 Department of Computer Science and Software Engineering, Auburn University, Auburn, AL, USA

*Address all correspondence to: galina.yakubova@usda.gov

References

[1] Wielopolski L. Nuclear methodology for non-destructive multi-elemental analysis of large volumes of soil. In: Carayannis E, editor. Planet Earth: Global Warming Challenges and Opportunities for Policy and Practice. London, UK: IntechOpen; 2011. pp. 467-492. DOI: 10.5772/23230

[2] Lugato E, Bampa F, Panagos P, Montanarella L, Jones A. Potential Carbon Sequestration of European Arable Soils Estimated by Modelling a Comprehensive Set of Management Practices. Global Change Biology. 2015; **20**:3557-3567. DOI: 10.1111/gcb.12551

[3] Barzilov AP, Novikov IS, Womble PC. Material analysis using characteristic gamma rays induced by neutrons. In: Adrovic F, editor. Gamma Radiation. London, UK: IntechOpen; 2012. pp. 17-40. Available from: http://www.intechopen.com/books/gamma-radiation/material-analysis-usingcharacteristic-gamma-rays-induced-by-pulse-neutrons. DOI: 10.5772/2054

[4] Kavetskiy A, Yakubova G, Torbert HA, Prior SA. Neutron-stimulated gamma ray analysis of soil. In: Maghraby AM, editor. New Insights on Gamma Rays. London, UK: IntechOpen; 2017. pp. 133-178. DOI: 10.5772/68014

[5] Yakubova G, Kavetskiy A, Prior SA, Torbert HA. Benchmarking the inelastic neutron scattering soil carbon method. Vadose Zone Journal. 2016;**15**(2):1-12. DOI: 10.2136/vzj2015.04.0056

[6] Yakubova G, Wielopolski L, Kavetskiy A, Torbert HA, Prior SA. Field testing a mobile inelastic neutron scattering system to measure soil carbon. Soil Science. 2014;**179**:529-535. DOI: 10.1097/SS.0000000000000099

[7] Yakubova G, Kavetskiy A, Prior SA, Torbert HA. Measurements of soil carbon by neutron-gamma analysis in static and scanning modes. Journal of Visualized Experiments. 2017;**126**: e56270. DOI: 10.3791/56270

[8] WaveMetrics. IGOR Pro [Internet]. 2017. Available from: http://www.wavemetrics.com/products/igorpro/ig orpro.htm [Accessed: December 01, 2021]

[9] ArcMap Desktop [Internet]. 2018. Available from: https://support.esri.com/en/products/desktop/arcgis-desktop/arcmap/10-5 [Accessed: December 01, 2021]

[10] Geant4, version 10.01, patch 2. [Internet]. 2015. Available from: https://geant4.web.cern.ch/support/download_archive?page=5 [Accessed: December 01, 2021]

[11] Kavetskiy A, Yakubova G, Sargsyan N, Wikle C, Prior SA, Torbert HA, et al. Scanning mode application of neutron-gamma analysis for soil carbon mapping. Pedosphere. 2019;**29**(3):334-343. DOI: 10.1016/S1002-0160

[12] ASTM D5922-18. 2018. Standard Guide for Analysis, Interpretation, and Modeling of Spatial Variation in Geostatistical Site Investigations. Available from: https://www.astm.org/d5922-18.html [Accessed: December 01, 2021]

[13] Windows Presentation Foundation. 2021. Available from: https://docs.microsoft.com/en-us/visualstudio/designers/getting-started-with-wpf?view=vs-2022 [Accessed: December 01, 2021]

[14] National Marine Electronics Association. 2021. Available from:

https://www.nmea.org/content/STANDARDS/NMEA_0183_Standard [Accessed: December 01, 2021]

[15] NMEA Revealed. 2021. Available from: https://gpsd.gitlab.io/gpsd/NMEA.html [Accessed: December 01, 2021]

[16] Develop Android apps with Kotlin. 2020. Available from: https://developer.android.com/kotlin [Accessed: December 01, 2021]

Chapter 6

An Overview of Soil Moisture and Salinity Sensors for Digital Agriculture Applications

Redmond R. Shamshiri, Siva K. Balasundram, Abdullah Kaviani Rad, Muhammad Sultan and Ibrahim A. Hameed

Abstract

Soil salinity and the water crisis are imposing significant challenges to more than 100 countries as dominant factors of agricultural productivity decline. Given the rising trend of climate change and the need to increase agricultural production, it is crucial to execute appropriate management strategies in farmlands to address salinity and water deficiencies. Ground-based soil moisture and salinity sensors, as well as remote sensing technologies in satellites and unmanned aerial vehicles, which can be used for large-scale soil mapping with high accuracy, play a pivotal role in precision agriculture as advantageous soil condition monitoring instruments. Several barriers, such as expensive rates and a lack of systematic networks, may hinder or even adversely impact the progression of agricultural digitalization. As a result, integrating proximal equipment with remote sensing and Internet of things (IoT) capabilities has been shown to be a promising approach to improving soil monitoring reliability and efficiency. This chapter is an attempt to describe the pros and cons of various soil sensors, with the objective of promoting IoT technology in digital agriculture and smart farming.

Keywords: precision agriculture, digital, soil sensors, moisture, salinity

1. Introduction

Drought and soil salinity are two of the world's dominant abiotic stresses that severely restrict crop production, and it is expected that these challenges, along with accelerating climate change, will drive universal food insecurity [1]. In parallel, the 933 million people affected by the water crisis in 2016 are expected to increase to 1.693–2.373 billion people in 2050 [2] as a consequence of the global increasing population and an additional rise in water demand [3]. Despite the fact that agriculture receives more than 70% of water supplies [4], most governments lack precise irrigation water usage statistics [5]. Irrigation processes waste 25–30% of fresh water, resulting in a loss of $14 billion. Therefore, proper water management is critical

[6, 7]. Otherwise, growers are compelled to use saline water for irrigation owing to water shortages that lead to soil salinity expanding [8]. Soil salinity is one of the most damaging agents to cropland in more than 100 countries [9, 10]. Salinity affects more than 25% of the world's terrestrial lands and a third of the world's irrigated fields [11]. The total area of saline soils is reported to be 1060.1 million hectares, with climate change driving this estimate to rise [12].

The factors that cause natural or primary salinity include parent materials and saline minerals in the soil. Anthropogenic factors, such as conventional irrigation techniques and weak drainage systems, cause secondary salinity [13]. Complications of the accumulation of excess soluble salts, specifically chloride sulfate [14] in the root zone of plants [15], include reducing plant growth, groundwater pollution, and diminishing soil fertility, ultimately degrading farmlands [11, 16]. High soil salinity decreases crop productivity, especially vegetables, which are extremely sensitive during the ontogeny stage. The salinity tolerance of most vegetables is low [17]. Castanheira et al. [18] observed that along with increasing the salinity of irrigation water to 5 $ds.m^{-1}$, the average solute concentration in the root zone reaches a level higher than the corn tolerance. Moreover, high salinity negatively impacts the physicochemical and biological traits of soils, such as the diversity and abundance of microbes and animals [19], consequently leading to adverse consequences for farmers' livelihoods, and the regional and national economy [20]. The financial loss caused by salinity-induced land degradation in 2013 was estimated at $441 per hectare, equivalent to $27 billion annually [21]. Hence, improper water management and subsequent salinization threaten the sustainability of agriculture [22]. Many investigations have been carried out to cope with the obstacles of water deficit and salinity. Irrigation water management strategies and drainage techniques as the most prevalent solutions [23, 24], specifically in arid and semiarid regions, can face numerous challenges such as high costs and inefficiency [11, 19, 25]. Notwithstanding investments in countering the salinity spread, farmers are still challenged by the consequences of soil salinity [26]. Food security is threatened whenever efficient management actions are not exerted to maintain agricultural production [27]. **Figure 1** shows the salinity and water stress situations in various regions of the world.

The uninterrupted monitoring of soil moisture and salinity in agriculture is accepted in order to limit water and salinity crises. After sea level temperature, soil moisture as a significant climatic determinant is the second prominent factor influencing evapotranspiration, sensible surface heat, and latent heat flux, as well as water, carbon, and energy cycles on a global scale [30, 31]. Changes in soil moisture alter both agricultural and municipal soils [32]. This essential variable is employed in order to improve weather forecasting, rainfall estimation, drought monitoring, and landslide and flood prediction [31]. There are multiple methods to measure soil moisture, which is directly correlated with irrigation efficiency [33]. Indirect methods estimate soil moisture using a gravimetric, gamma-radiation probe, neutron probe, and porous blocks based on gravitational sampling or time-domain reflectometry (TDR) in a small soil bulk. Direct methods also evaluate soil moisture using weighted moisture *in vitro* [34, 35]. In most circumstances, soil moisture is not directly measurable; instead, it is measured indirectly through moisture-related characteristics [36]. TDR is extensively employed to identify the soil water content according to the connection between dielectric constant and moisture content [34]. However, a study in the USA ascertained that only 1.2 out of 10 farms use soil moisture sensors for irrigation planning. This quantity is lower globally due to a lack of systematic support, sensor inconsistency, and high costs, resulting in the rejection of these systems [37].

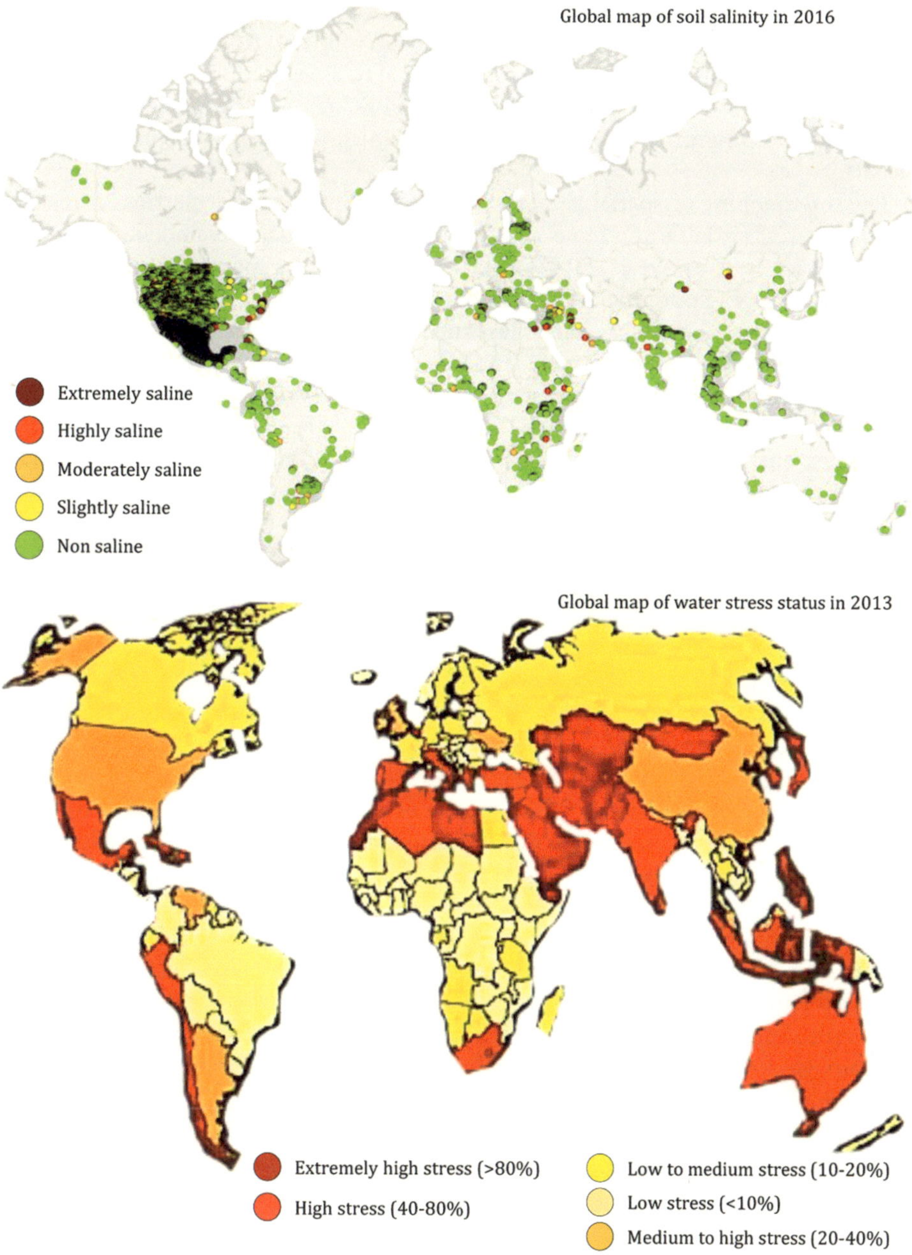

Figure 1.
Map of global soil salinity and water stress status. Adapted from [28, 29].

A thorough understanding of soil salinization processes is also required for long-term soil and water management [38], which employs conventional electrical conductivity (EC) sensors [39]. In addition to salinity, EC is an indicator of soil health and nutrient availability for plants [40]. Salinity sensors are designed according to three electromagnetic (EM) phenomena: (i) electrical resistance, (ii) electromagnetic induction, and (iii) reflectometry [41]. The most accurate commercial method of EC estimation is the application of electromagnetic induction, including four electrodes [42]. EM38 is a noninvasive soil electromagnetic induction sensor that can measure EC at 120 cm above the soil and assess the soil nutrient situation [43, 44]. Although soil salinity modeling in farmlands using EC sensors is crucial to assess crop yield and prevent productive soil loss [45], measuring apparent soil EC (ECa) is needed for calibration with the actual content of salts in the laboratory [46], which is not economically cost-effective.

The soil mapping of spatial and temporal variations in soil properties is presumably the most affordable and beneficial approach to front salinity and watering issues. In this regard, Mashimbye et al. [47] evaluated the role of hyperspectral or satellite data in soil mapping potential applications. Satellite technologies make it easier to measure salinity and moisture variables, and as a result, they can provide soil characteristic data instantly, quantitatively, and affordably [48]. For instance, the launching of Sentinel satellites upgraded free data access for users [49], including advanced facilities for earth monitoring [50]. Though the remote sensing of soil properties presents extensive coverage for spatial distribution, multispectral data have limited capabilities, such as low spatial resolution due to spectral and spatial division [35, 50]. A spatial description of soil salinity is essential for salinity management in agriculture [51]. On the other hand, conventional techniques for evaluating soil characteristics are costly and time-consuming [52] (**Figure 2**); the question of whether proximal sensors or aerial sensors are more efficient for controlling soil moisture and salinity levels in farmlands arises.

2. Proximal soil sensing

Facing the growing demand for food and sustaining water resources needs irrigation optimization employing advanced technologies such as soil moisture sensors [53]. Technologies such as drip watering, proximal sensors, and remote controllers for water management have joined the farming sector owing to agricultural development and subsequently rising demand for freshwater [54]. Considering that implementing a systematic irrigation plan for farmers is practically complicated, digital instruments effectively assist in accurate irrigation planning [55]. Furthermore, the proximal platform can be used to evaluate plant health [56]. Recent advances in electromagnetic moisture sensor technologies have facilitated automatic irrigation

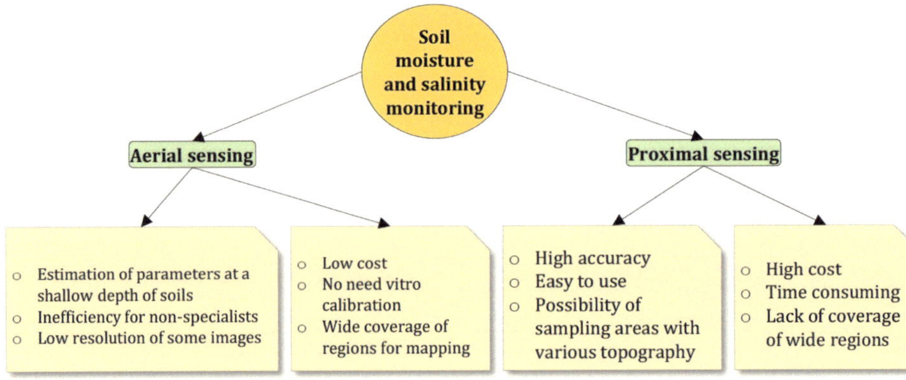

Figure 2.
Positive and negative attributes of proximal and aerial sensors.

scheduling [57], which enhances water-use efficiency. These sensors are divided into active and passive instruments, which are applied for crop yield assessment and watershed management in digital agriculture [58]. In another classification system, soil sensors can be divided into resistive or capacitive sensors. Resistance-based sensors are easy to use and inexpensive. However, error sources affect their accuracy and efficiency [59]. Jusoh et al. [60] reported that the resistive sensor operates defectively in sandy loam and clay loam soils owing to low bulk density and high organic matter.

As efficient machines, capacitive soil moisture sensors are affordable for reducing water costs and wastage and computerized scheduling of irrigation [57, 61]. Capacitive probes and electronic TDR soil moisture sensors with *in situ* measurement have easy use, high accuracy, and fast data retrieval that are extensively used to monitor soil moisture changes in fields and predict drought, particularly in arid and semiarid lands. Furthermore, these instruments are applied for hydrological flux calculations, modeling runoff infiltration, and calibration of remote sensing data. However, precisely estimating moisture content is not convenient due to the spatial diversity of soils and the spatiotemporal heterogeneity of soil water content at high depths [36]. It is further challenging to measure moisture using discrete and wire-based instruments in fields with high vegetation diversity and different hydrological properties, which cause numerous obstacles in analysis and control systems, specifically at broad geographical scales [55, 62]. Since some sensors retrieve various data from a farm, it is not possible to automatically turn on or off the federal irrigation system. Moreover, many users have reported fractures of the watermark rod during dipping or separating it from the soil (**Figure 3**). The low accuracy of some sensors, which have a high moisture detection limit and detect the soil as dry, directly challenged farmers. Therefore, there is a possibility of flooding the root zone and loss of plants in the event of inadequate knowledge of farmers. Hence, growers' propensity to purchase sensors decreases. The cost of sensors determines their resistance and precision in heterogeneous ambient conditions [63]. A flawless calibration process is necessary to optimize the sensor's accuracy. In order to improve the accuracy of the soil moisture sensor, Gonzalez-Teruel et al. [64] calibrated it on three different types of soil. According to Radi et al. [65], the soil moisture sensor SKU:SEN0193 is a low-cost commercial sensor that must be calibrated before being used on farms. **Figure 4** shows the calibration process of a soil moisture sensor. Since different raw materials are used to make sensors, low-cost sensors have low resistance to

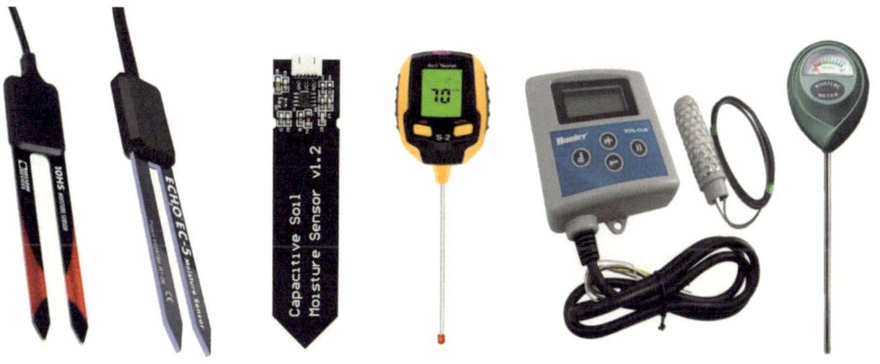

Figure 3.
Instances of different soil moisture sensor probes that are used for digital farming applications.

Figure 4.
Calibration of soil moisture sensor for different types of soil. Source: SunBot.de.

adverse environmental conditions such as sunlight, strong winds, and wild animals. Therefore, it is challenging to achieve integrated systems on farms owing to the natural obstacles. A proximal network is high-priced due to the need for periodic servicing of sensor portions [66], which increases costs for producers. Given that experimental determination of soil moisture is a fundamental characteristic of agricultural operations [66, 67], cost-effective analysis of soil volumetric water content (VWC) is an important strategy for promoting sustainable agriculture through the use of computerized machines and Internet of things (IoT) development, particularly for smallholder farmers [68].

Significant advances have been made in technologies for assessing, mapping, and spatiotemporal monitoring of salinity on a field, regional, and national scale [10]. Generally, there are five methods for estimating salinity on a farm: (1) observing salts on the soil surface, (2) estimating EC in saturated soil extracts, (3) measuring *in situ* electrical resistance, (4) determining *in situ* EC by TDR, and (5) noninvasive EC measurement using EM sensors [69]. The EM38 sensor is one of the most popular sensors in agriculture and consists of a receiver and a transmitter coil with a distance of 1 meter from each other, which are connected at the opposite end of a nonconducting rod, which measures salinity and other soil properties such as nutrient level and clay bulk [70]. This sensor is comfortable to use, and users can interpret its data after processing obtained images [71]. Slavich et al. [72] and Guo et al. [73] used EM38 data to determine soil salinity and barley tolerance to salinity and for digital soil mapping of spatiotemporal salinity changes. Hammam and Mohamed [74] mapped the spatial pattern of soil salinity in the East Nile Delta using geographic information system (GIS) and inverse distance weighting (IDW) techniques. Ding and Yu [75] reported that the obtained EC data from the EM38 sensor were significantly correlated with the experimental soil analysis in the laboratory. Guo et al. [76] recognized a significant correspondence between actual soil EC and sensor data ($r > 0.9$) by employing EM38 proximal technology. Additionally, EM38 is beneficial for prompt soil assessment before planting operations (**Figure 5**) [77]. Despite the speedy operation of this sensor, its vulnerability to metals and electromagnetic noise sources, such as power cables, can generate fluctuations in data registration [78]. The framework of the soil moisture proximal measuring procedure is outlined in **Figure 6**, along with three models of soil sampler robots.

The integrated wireless sensor network (WSN) is designed to measure soil salinity and support automated irrigation systems [83]. With the WSN, numerous facilities are provided such as remote monitoring of soil fertility, crop water situation, and assistance to the irrigation system with reasonable costs, low energy consumption,

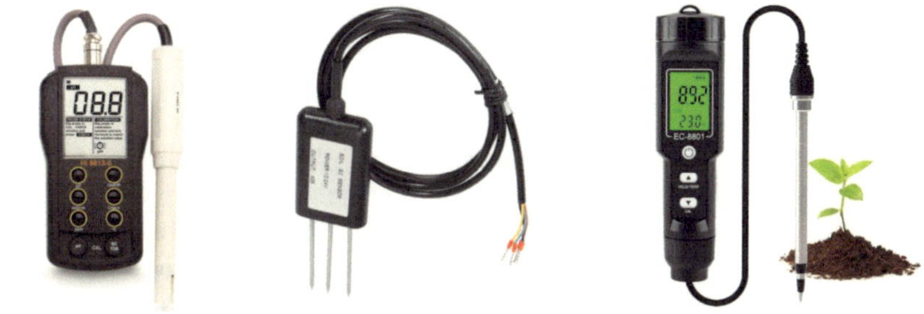

HI9813-6 Portable pH, EC, TDS, and Temperature Meter
Source: hannainst.com.au

RS485 waterproof soil salinity ec fc sensor transmitter soil conductivity Det BS
Source: Ita Nest

EC-8801BB portable soil EC temperature test meter with detachable probe
Source: Yieryi

Figure 5.
Instances of different portable pH and EC meters used for measuring soil salinity.

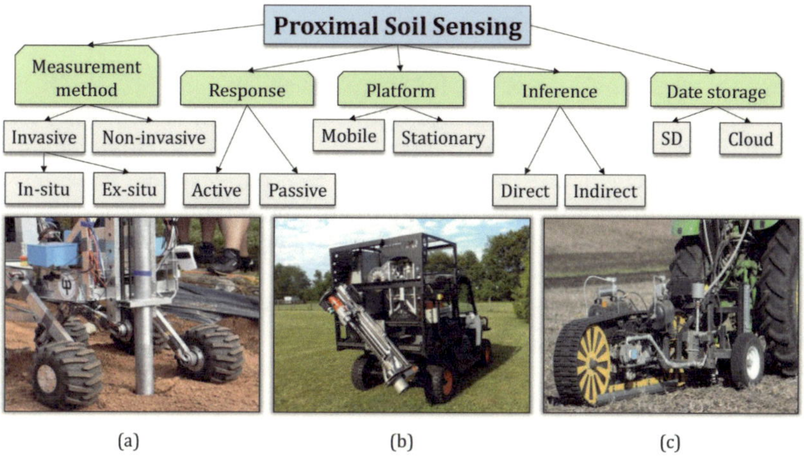

Figure 6.
Summary of the soil moisture measurement process by the proximal sensor and models of soil sampler robots. Adapted from [79–82].

and extended life [84, 85]. In a study by Sui and Baggard [86], WSN sensors automatically recorded soil condition data over the Internet every minute. The combination of WSN with the GIS in a study by Zhang et al. [87] proposed a soil moisture distribution map for accurate irrigation control. This system improves irrigation efficiency by decreasing freshwater loss and watering costs [88]. The precision of the data retrieved by the WSN depends on the system's capability to hold the input voltage constant and the dependability of the calibration curves [89]. Though the WSN with fast data retrieval capability is a promising strategy in precision agriculture, barriers such as soil and canopy interference can affect data validity [79].

3. Aerial soil sensing

3.1 Drone-based remote sensing

Monitoring soil conditions with remote sensing systems is a new approach that enhances productivity in digital agriculture [90]. Through the development of unmanned aerial vehicle (UAV) technologies [91, 92], it is now possible to retrieve soil property data with high resolution and low cost for mapping. UAVs reliably transfer soil characteristic data to computers, thereby playing an important role in precision agriculture [93]. When compared to satellites, UAVs have superior control and high spatial resolution [94]. Hu et al. [95] reported that UAVs using 62 hyperspectral bands afforded more reliable data for soil salinity prediction models than satellites, making UAVs a valuable machine for small-scale soil mapping. UAVs are also useful for assessing soil moisture in heterogeneous landscapes [96]. In addition to soil moisture, multispectral images of UAVs can be applied to map the distribution of water stress in crops (**Figure** 7) [98]. Although UAVs play a prominent role in precision agriculture, further attempts should be made to derive from data processing techniques and vegetation calibration in the future [99]. Moreover, UAVs face other challenges, such as limited flight time and stabilization, so future studies should concentrate on addressing these problems [100].

3.2 Satellite-based remote sensing

High spatial resolution is necessary for analyzing soil moisture [101]. Thereby, satellites are the principal instruments for characterization and monitoring soil moisture with an accuracy of approximately 5 cm [102]. Ahlmer et al. [103] demonstrated that using satellite data enhances the reliability of flood forecasting. The microwave brightness temperature is sensitive to soil moisture content due to water affecting the dielectric constant [104]. In recent years, digital agriculture has made enormous progress in estimating soil moisture by applying microwave sensors. In contrast, advancements have been restrained owing to heterogeneities between satellite data resolution and hydrological scales, vegetation, and low microwave infiltration [105]. Satellite

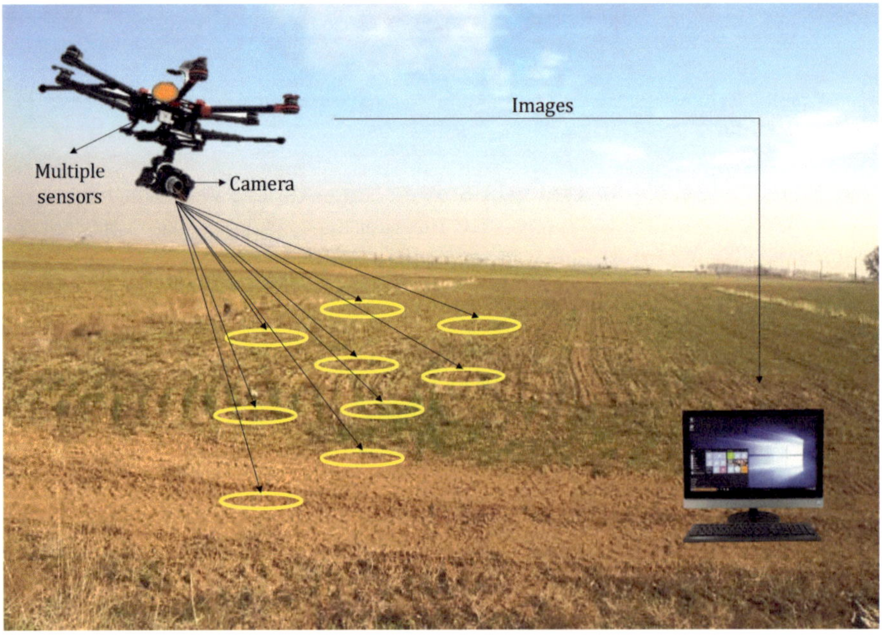

Figure 7.
The measurement process of soil moisture using UAVs. Adapted from [97].

sensors are potentially designed to monitor vast regions; however, their spatial resolution depends on the microwave frequency, antenna size, and elevation. Most passive radiometers have a spatial resolution of 10 km, which is inapplicable for hydrological aims. Although microwave remote sensing drives many algorithms for calculating large-scale soil moisture, their low resolution is not appropriate for small scale [106]. Presently, the passive microwave retrieved resolution of soil moisture is about 25 km [107], and the low spatial resolution outputs, unreliable rainfall, and evaporation-transpiration data can make it challenging to estimate irrigation water demand [5]. Moreover, soil moisture data may not be available regularly. The spatial distribution of soil moisture is a prerequisite for agricultural and ecological management, while retrieving soil moisture data in heterogeneous landscapes is a significant challenge [108]. Heterogeneous landscapes generate irregularities in moisture measurements [109]. Consequently, merging surface reflectance data and auxiliary geospatial data can accurately estimate soil moisture, supporting precision agriculture strategies efficiently. **Table 1** summarizes some investigations that measured soil moisture using a combination of proximal and satellite data.

Satellite	Application	Location	Result	Reference
ASCAT	Estimating soil moisture	Arizona (USA)	The geostatistical approach is beneficial to estimate soil moisture for network cells without data from satellite imagery.	[110]
Envisat	Hydrological modeling	Okavango (Southern Africa)	Remote sensing improves the hydrological model for unsuccessfully evaluated watersheds.	[111]
MODIS	Estimating soil moisture	Henan (China)	Applying meteorological data to missing pixels of the satellite can enhance the accuracy of estimation and afford a comprehensive map of soil moisture in broad regions.	[112]
Landsat	Mapping water consumption	Tensift Al Haouz (Morocco)	There is a correlation between the satellite NDVI index, soil evaporation, and cover fraction variables.	[113]
SMOS	Assessing soil moisture for drought monitoring	Iran	It was reported that the central and southeastern regions had experienced the most severe drought in 2000–2014.	[114]
MODIS	Monitoring soil and vegetation moisture	Kansas and Oklahoma (USA)	The drought sensitivity was significantly improved by combining several infrared bands of the satellite.	[115]
SMOS	Monitoring drought for agricultural purposes	Remedhus (Spain)	SWDI reflects the soil water balance dynamics and can monitor drought in agriculture.	[116]
MODIS	Monitoring drought for agricultural purposes	Korean peninsula	The High-resolution Soil Moisture Drought Index (HSMDI) was significantly correlated with crop yield data.	[117]

Table 1.
Some studies on the merged application of ground-based and satellite sensors to estimate soil moisture.

Remote sensing data can be applied to map surface soil salinity in broad regions [39], and the Landsat satellite has made it attainable to study soil salinity at different scales [118]. Wu et al. [119] reported that the overall accuracy of Landsat in soil salinity detection from 1973 to 2006 was approximately 90.2%. Combining proximal instruments with remote sensing systems is advantageous in precision evaluating soil salinity [120]. Bouaziz et al. [121] extracted 18 indicators from MODIS Terra data to improve salinity prediction patterns in northeastern Brazil and recognized a moderate correlation between EC and spectral indices. However, the limitations of using remote sensing data to map salt-affected areas include salt spatial distribution, temporal changes, and vegetation interference [122]. Moreover, it is challenging to estimate soil salinity through single-factor models [123]. Although remote sensing has numerous advantages over conventional proximal systems for mapping and predicting soil salinity [124], it is possible to determine the spatial variability of soil EC by local proximal sensor EM38 connected to GPS [125]. Casterad et al. [126] applied a combination of soil experiment data, proximal sensors, and satellites to investigate how soil salinity develops and distributes. Corwin [127] used proximal sensors and remote imaging to assess soil salinity at different scales; furthermore, Douaoui et al. [128] demonstrated that the regression-kriging approach combines remote sensing systems and ground network monitoring stations, thereby providing well-defined spatial and temporal monitoring of soil salinity. Eldeiry and Garcia [129] similarly reported that the modified kriging model presents the most reliable estimate of soil salinity by combining satellite and proximal data.

In a study by Fourati et al. [52], ordinary kriging with an average of 1.83 squares and a standard error of 0.018 had the most reliable performance for identifying and classifying saline soils. In an investigation by Fan et al. [130], the partial least squares regression model was applied to retrieve soil salinity from multispectral sensors, allowing salinity mapping with low cost and significant accuracy. Yahiaoui et al. [131] analyzed the topographic characteristics of the study area using Landsat 7 satellite data; accordingly, they created a multiple linear regression based on height and an adjusted soil salinity index that could predict soil salinity by 45%. Soil salinity modeling by satellite and proximal data in central Iraq revealed that models could reliably forecast salinity with 82.57% precision [119]. Therefore, modeling spatial soil salinity changes based on remote sensing data regression analysis is an economical, simple, and promising approach [132].

4. Wireless sensing and IoT monitoring

The precision agriculture approach employs new technologies to optimize farming inputs and ameliorate agricultural systems [133]. As one of the newest Internet-based technologies to have joined the agricultural sector, IoT is a type of intelligent sensor with software based on a web connection, applied to proposed purposes on farms. It drives modern agriculture toward the automatization of manual operations [134, 135], and its architecture is shown in **Figure 8**. The Wi-Fi module forwards the soil parameter data assembled by the sensors to the controller and processor [136]. Growers can inspect soil moisture, temperature, and pH data on an Android mobile phone using IoT technology [137]. Automated irrigation can also minimize human mediation [138] as an incentive to save more water [139]. Yamin et al. [140] demonstrated that a digital soil test kit connected to the IoT system could be used to dynamically evaluate changes in soil elements. Moreover, IoT can help optimally control

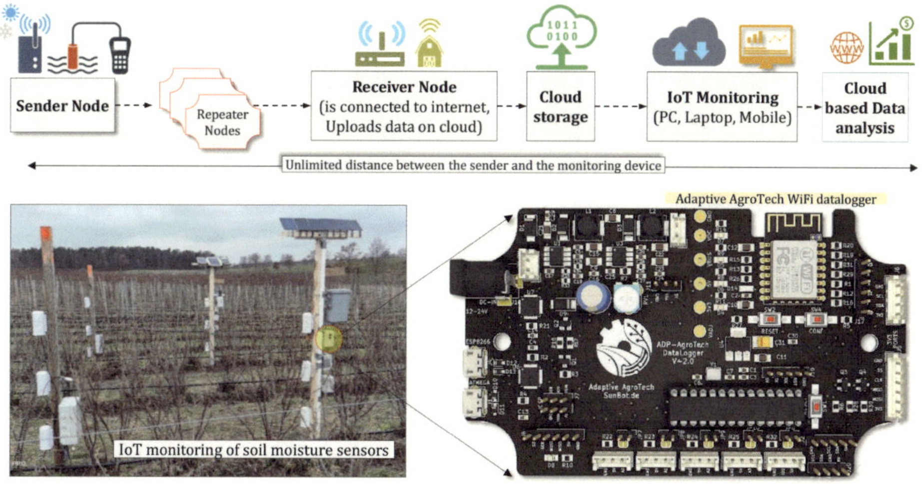

Figure 8.
Deployment of hybrid data logger with Wi-Fi connectivity for IoT monitoring of soil moisture in berry fields. Source: SunBot.de.

Figure 9.
Wireless monitoring of soil moisture with solar-powered modular sensors.

greenhouse conditions [141]. Shamshiri et al. [142] applied a systematic approach to automatically retrieving and processing greenhouse condition data in order to enhance tomato yield. Divyavani and Rao [143] could receive moisture sensor data using the Android mobile phone. Payero et al. [144] controlled soil moisture in a field through a mobile-based IoT system. The WSN system proposed by Shylaja and Veena [85] dispatched soil fertility circumstances to the mobile phone that are beneficial for fertilizer recommendation.

Figure 9 demonstrates a solar-powered hybrid (Wi-Fi, LoRa, data logger) soil moisture and salinity sensors that were deployed in commercial berry fields in Germany. This device benefits from an onboard memory module for logging the measurements before transmitting the data via Wi-Fi and LoRa. It should be noted that due to the rising salinity trend caused by climate change, these devices are required for the precision monitoring of soil salinity in small and large scales [128]. Evaluating salinity-affected zones combats global climate change and prevents water resource loss [145]. Soil mapping is crucial for determining positional salinity levels and promoting appropriate management strategies for saline land restoration [146]. Therefore, combining remote sensing systems and EM38 sensors has provided an accurate soil salinity assessment approach, which is necessary to prevent further land salinization [76]. Future studies should concentrate on advancing remote sensing technologies for soil properties and the integration of salinity maps [147]. The measurement of soil moisture is critical in predicting drought and warning of natural disasters. Recently, many attempts have been made to address the development of soil moisture measurement facilities [148]. Launching advanced satellites promotes new innovative research approaches and encourages the development of new systematic empirical techniques for measuring soil moisture [149]. Non-cost-effectiveness plus inaccessibility to soil characteristics is one of the most significant constraints of precision agriculture [150]. Future soil moisture sensors should have high precision, low cost, and nondestructive features. Prospective research should also include the creation of specialized sensors for specific situations [33]. Using soil probes is critical for the most efficient and cost-effective use of water and chemical fertilizers; thus, numerous experiments on soil health indicators, such as water-holding capacity, salinity, temperature, pH, and soluble gas concentrations, are carried out [151]. However, high costs and the complex protection of sensors prevent the development of digital farming technologies, especially in rural regions [152].

For the purpose of downloading data from multiple sensors, a standalone software application shown in **Figure 10** was developed by Adaptive AgroTech to interface with the sensors' controllers via multiple serial COM ports as well as to execute commands and set custom configurations. The software also provides users with other features such as downloading log files of the sensor performance (i.e., battery and clock status, or historical parameters) or uploading the stored data to a cloud server. In addition, users can set labels to each node for simultaneously reading and writing log files from multiple devices and store the data on local memory cards. The Adaptive AgroTech Port Logger was developed in C# programming language environment and the Microsoft dot Net Core technology and can be operated on Microsoft Windows, Apple macOS, and Linux operating system. It should be noted that the MS-DotNet is a free open-source software for cross-platform development that supports various languages, such as C#, C++, and VB.NET. These features have provided a cost-effective and flexible solution for the future improvement of the Port logger. To have the best result and optimum performance, the software uses multi-threading technology to execute parallel routines such as listening to multiports and executing more than one task at a time. Each thread defines a unique flow of control. As soon as the port logger engages in complicated and time-consuming parallel operations, it automatically sets different execution paths or threads, with each thread performing a particular task.

For the purpose of a visual comparison between air temperature, soil temperature, and soil surface moisture, sample data from the hybrid data logger shown in **Figure 9** that were collected every 10 minutes for 13 days in March 2021 are plotted in **Figure 11**. These plots validate the sensitivity of the sensor for the continuous monitoring

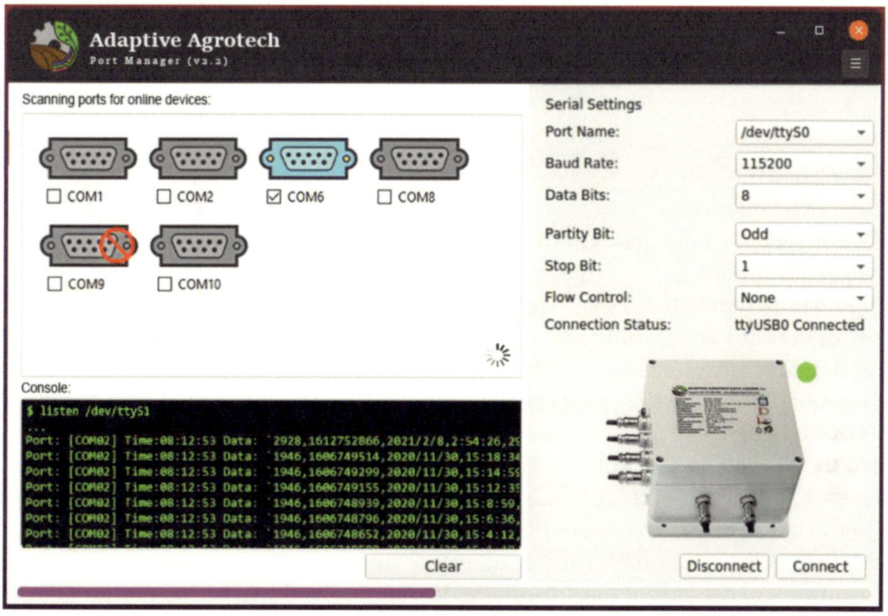

Figure 10.
Adaptive AgroTech Port Logger software for simultaneously downloading data from multiple sensor nodes under windows and Linux operating system.

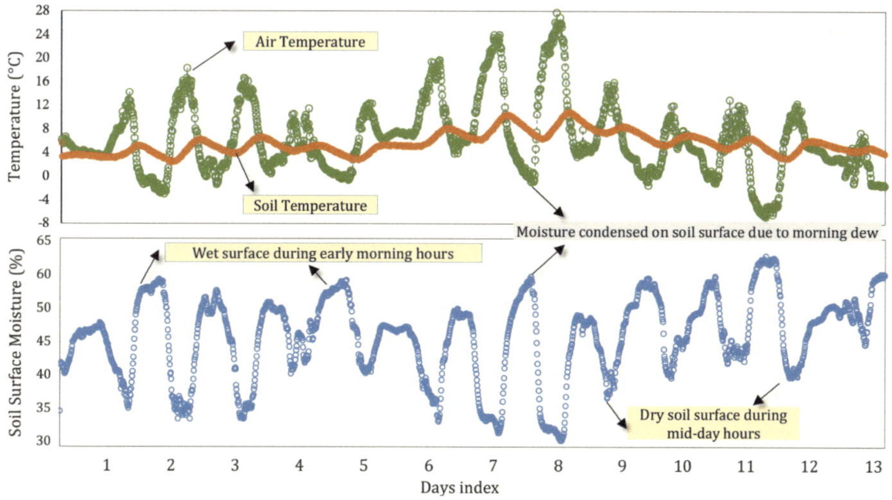

Figure 11.
Plots of air temperature, soil temperature, and soil surface moisture during 13 days of experiment for performance evaluation of an adaptive AgroTech hybrid data logger.

of agricultural field and for planning precision irrigation practices in arid areas. The measurements of the hybrid data logger can be used as the feedback data for a decision support system or controller that activates the irrigation pumps based on air and soil temperature, soil moisture, hours of the day, and other field parameters. It can be seen from the plots of **Figure 11** that during early morning hours, soil surface experiences more moisture (due to the morning dew) in the entire 13 days of the experiments compared to the mid-day hours. It can also be seen that the hybrid datalogger did not miss a single measurement during the experiments, even when the air temperature was below the freezing point.

5. Conclusion

As two global challenges without national borders, soil salinity and the water crisis endanger sustainable agricultural production through decreasing farmland productivity and crop yield [153, 154]. These principal abiotic stresses significantly restrict crop productivity by inhibiting metabolic activities and disturbing the ionic balance. Water deficits caused by osmotic stress severely reduce the crop yield, which drives considerable economic losses for farmers. Hence, monitoring their changes in farmlands using sensors is crucial due to the significant regional or national financial loss caused by drought and salinity. Despite soil moisture and salinity probes effectively measuring soil parameters, inefficient performance in broad fields plus the high cost and low accuracy have accelerated the application of new remote sensing technologies. Satellites and UAVs have the possibility of monitoring these variables on a broad scale. However, low spatial resolution, difficulty of use, the need for technological operators, and lengthy data processing make them unpopular with farmers, particularly in rural regions. In addition to remote sensing, IoT technology combines sensor systems and web-based software that transfers soil moisture and salinity data to a computer or mobile phone. While precision agriculture is gradually developing new technologies in farmlands, more extensive investigations are needed to address the challenges of agricultural digitalization.

Author details

Redmond R. Shamshiri[1*], Siva K. Balasundram[2], Abdullah Kaviani Rad[3], Muhammad Sultan[4] and Ibrahim A. Hameed[5]

1 Leibniz Institute for Agricultural Engineering and Bioeconomy, Potsdam, Germany

2 Department of Agriculture Technology, Universiti Putra Malaysia, Serdang, Selangor, Malaysia

3 Department of Soil Science, School of Agriculture, Shiraz University, Shiraz, Iran

4 Department of Agricultural Engineering, Bahauddin Zakariya University, Multan, Pakistan

5 Faculty of Information Technology and Electrical Engineering, Department of ICT and Natural Sciences, Norwegian University of Science and Technology (NTNU), Ålesund, Norway

*Address all correspondence to: rshamshiri@atb-potsdam.de

References

[1] Wen W, Timmermans J, Chen Q, van Bodegom PM. A review of remote sensing challenges for food security with respect to salinity and drought threats. Remote Sensing. 2021;**13**(1):1-14

[2] He C, Liu Z, Wu J, Pan X, Fang Z, Li J, et al. Future global urban water scarcity and potential solutions. Nature Communications. 2021;**12**(1):1-11

[3] FAO. Water Scarcity—One of the greatest challenges of our time. Accessed date: 20 March 2019. Available from: https://www.fao.org/fao-stories/article/en/c/1185405/

[4] United Nations. Summary Progress Update 2021: SDG 6 — water and sanitation for all. Accessed date: 24 February 2021. Available from: https://www.unwater.org/publications/summary-progress-update-2021-sdg-6-water-and-sanitation-for-all/

[5] Jalilvand E, Tajrishy M, Ghazi Zadeh Hashemi SA, Brocca L. Quantification of irrigation water using remote sensing of soil moisture in a semi-arid region. Remote Sensing of Environment. 2019;**231**:111226

[6] United Nations. Water scarcity. Accessed date: 20 November 2021. Available from: https://www.unwater.org/water-facts/scarcity/

[7] Taft HL. Water Scarcity: Global Challenges for Agriculture. Food, Energy, and Water: The Chemistry Connection. Netherlands: Elsevier Inc; 2015. pp. 395-429

[8] Helalia SA, Anderson RG, Skaggs TH, Šimůnek J. Impact of drought and changing water sources on water use and soil salinity of almond and pistachio orchards: 2. modeling. Soil Systems. 2021;**5**(4):58

[9] Boudibi S, Sakaa B, Benguega Z, Fadlaoui H, Othman T, Bouzidi N. Spatial prediction and modeling of soil salinity using simple cokriging, artificial neural networks, and support vector machines in El Outaya plain, Biskra, southeastern Algeria. Acta Geochimica. 2021;**40**(3):390-408

[10] Shahid SA. Developments in Soil Salinity Assessment, Modeling, Mapping, and Monitoring from Regional to Submicroscopic Scales. In: Shahid S, Abdelfattah M, Taha F, editors. Developments in Soil Salinity Assessment and Reclamation. Dordrecht: Springer; 2013

[11] Mohanavelu A, Naganna SR, Al-Ansari N. Irrigation induced salinity and sodicity hazards on soil and groundwater: An overview of its causes, impacts and mitigation strategies. Agriculture (Switzerland). 2021;**11**(10):983

[12] Eswar D, Karuppusamy R, Chellamuthu S. Drivers of soil salinity and their correlation with climate change. Current Opinion in Environmental Sustainability. 2021;**50**:310-318

[13] Gorji T, Sertel E, Tanik A. Monitoring soil salinity via remote sensing technology under data scarce conditions: A case study from Turkey. Ecological Indicators. 2017;**74**:384-391

[14] Toderich KN et al. Integrating Agroforestry and Pastures for Soil Salinity Management in Dryland Ecosystems in Aral Sea Basin. In: Shahid S, Abdelfattah M, Taha F, editors. Developments in Soil Salinity Assessment and Reclamation. Dordrecht: Springer; 2013

[15] Yurtseven E, Öztürk H, Sertan A. Mass balance criteria in soil salinity management: The effect of different irrigation water qualities and leaching ratio. Journal of Agricultural Sciences. 2014;**20**(2):103-111

[16] Kulmatov R, Khasanov S, Odilov S, Li F. Assessment of the space-time dynamics of soil salinity in irrigated areas under climate change: A Case Study in Sirdarya Province, Uzbekistan. Water, Air, and Soil Pollution. 2021;**232**(5):216

[17] Machado RMA, Serralheiro RP. Soil salinity: Effect on vegetable crop growth. Management practices to prevent and mitigate soil salinization. Horticulturae. 2017;**3**(2):30

[18] Castanheira NL, Paz AM, Farzamian M, Paz MC, Santos FM, Fernandes ML, et al. Modelling of soil water and salt dynamics and prediction of salinity risks in Lezíria (Portugal) in response to different irrigation water qualities. Revista de Ciências Agrárias. 2020;**43**(2):161-173

[19] Sahab S, Suhani I, Srivastava V, Chauhan PS, Singh RP, Prasad V. Potential risk assessment of soil salinity to agroecosystem sustainability: Current status and management strategies. Science of the Total Environment. 2021;**764**:144-164

[20] Gopalakrishnan T, Kumar L. Soil Salinity to Paddy Land Abandonment in Jaffna Peninsula. 2021;211. 10.3390/agriculture

[21] Shahid SA, Zaman M, Heng L. Soil Salinity: Historical Perspectives and a World Overview of the Problem. Guideline for Salinity Assessment, Mitigation and Adaptation Using Nuclear and Related Techniques. 2018;43-53. 10.1007/978-3-319-96190-3_2

[22] Devkota M, Gupta RK, Martius C, Lamers JPA, Devkota KP, Sayre KD, et al. Soil salinity management on raised beds with different furrow irrigation modes in salt-affected lands. Agricultural Water Management. 2015;**152**:243-250. DOI: 10.1016/j.agwat.2015.01.013

[23] Ren D, Wei B, Xu X, Engel B, Li G, Huang Q, et al. Analyzing spatiotemporal characteristics of soil salinity in arid irrigated agro-ecosystems using integrated approaches. Geoderma. 2019;**356**:113935

[24] Based C, Tools S, Salinity FOR, Lands M, Goel A, et al. Review of computer based software tools for salinity management in agricultural lands. Journal of Indian Water Resources Society. 2013;**33**(4):24-32

[25] Panagea IS, Daliakopoulos IN, Tsanis IK, Schwilch G. Evaluation of promising technologies for soil salinity amelioration in Timpaki (Crete): A participatory approach. Solid Earth. 2016;**7**(1):177-190

[26] Wichelns D, Qadir M. Achieving sustainable irrigation requires effective management of salts, soil salinity, and shallow groundwater. Agricultural Water Management. 2015;**157**:31-38

[27] Zarei M, Rad AK. Covid-19. Challenges and Recommendations in Agriculture. 2020;**02**(01):12-15

[28] Ivushkin K, Bartholomeus H, Bregt AK, Pulatov A, Kempen B, de Sousa L. Global mapping of soil salinity change. Remote Sensing of Environment. 2019;**231**:111260

[29] Gassert F, Reig P, Luo T, Maddocks A. Aqueduct country and river basin rankings: a weighted aggregation of spatially distinct hydrological indicators. 2013;(December):28. Available

from: https://wri.org/publication/aqueduct-country-river-basin-rankings

[30] Shi CX, Xie ZH, Qian H, Liang ML, Yang XC. China land soil moisture EnKF data assimilation based on satellite remote sensing data. Science China Earth Sciences. 2011;**54**(9):1430-1440

[31] Rodríguez-Fernández N, Al Bitar A, Colliander A, Zhao T. Soil moisture remote sensing across scales. Remote Sensing. 2019;**11**(2):1-5

[32] Lo TH, Rudnick DR, Singh J, Nakabuye HN, Katimbo A, Heeren DM, et al. Field assessment of interreplicate variability from eight electromagnetic soil moisture sensors. Agricultural Water Management. 2020;**231**:105984

[33] Yu L, Gao W, Shamshiri RR, Tao S, Ren Y, Zhang Y, et al. Review of research progress on soil moisture sensor technology. International Journal of Agricultural and Biological Engineering. 2021;**14**(4):32-42

[34] Quinones H, Ruelle P, Nemeth I. Comparison of three calibration procedures for TDR soil moisture sensors. Irrigation and Drainage. 2003;**52**(3):203-217

[35] Fontanet M, Fernàndez-Garcia D, Ferrer F. The value of satellite remote sensing soil moisture data and the DISPATCH algorithm in irrigation fields. Hydrology and Earth System Sciences. 2018;**22**(11):5889-5900

[36] Paige GB, Keefer TO. Comparison of field performance of multiple soil moisture sensors in a semi-arid rangeland. Journal of the American Water Resources Association. 2008;**44**(1):121-135

[37] Kukal MS, Irmak S, Sharma K. Development and application of a performance and operational feasibility guide to facilitate adoption of soil moisture sensors. Sustainability (Switzerland). 2020;**12**(1):321

[38] Inam A, Adamowski J, Prasher S, Halbe J, Malard J, Albano R. Coupling of a distributed stakeholder-built system dynamics socio-economic model with SAHYSMOD for sustainable soil salinity management—Part 1: Model development. Journal of Hydrology. 2017;**551**:596-618

[39] Aldabaa AAA, Weindorf DC, Chakraborty S, Sharma A, Li B. Combination of proximal and remote sensing methods for rapid soil salinity quantification. Geoderma. 2015;**239**:34-46

[40] Arnold SL, Doran JW, Schepers J, Wienhold B, Ginting D, Amos B, et al. Portable probes to measure electrical conductivity and soil quality in the field. Communications in Soil Science and Plant Analysis. 2005;**36**(15-16):2271-2287

[41] Visconti F, de Paz JM. Electrical conductivity measurements in agriculture: The assessment of soil salinity. New Trends and Developments in Metrology (IntechOpen). 2016:99-126

[42] Rhoades J. Electrical conductivity methods for measuring and mapping soil salinity. Advances in agronomy. 1993;**49**:201-251

[43] Fritz RM, Malo DD, Schumacher TE, Clay DE, Carlson CG, Ellsbury MM, et al. Field Comparison of Two Soil Electrical Conductivity Measurement Systems. 2015:1211-1217

[44] Robinson DA, Lebron I, Lesch SM, Shouse P. Minimizing drift in electrical conductivity measurements in high temperature environments using the EM-38. Soil Science Society of America Journal. 2004;**68**(2):339-345

[45] Grisso R, Alley M, Holshouser D, Thomason W. Precision farming tools: Soil electrical conductivity. Virginia Cooperative Extension. 2009;**442**(508):1-6

[46] Hardie M, Doyle R. Chapter 28: Measuring Soil Salinity. In: Plant Salt Tolerance: Methods and Protocols, Methods in Molecular Biology (Methods and Protocols). Vol. 913. Totowa, NJ: Humana Press; 2018. pp. 415-425

[47] Mashimbye ZE, Cho MA, Nell JP. Model-based integrated methods for quantitative estimation of soil salinity from hyperspectral remote sensing data: A Case Study of Selected South African Soils. Pedosphere. 2012;**22**(5):640-649

[48] Gorji T, Yildirim A, Sertel E, TANIK A. Remote sensing approaches and mapping methods for monitoring soil salinity under different climate regimes. International Journal of Environment and Geoinformatics. 2019;**6**(1):33-49

[49] Ali I, Greifeneder F, Stamenkovic J, Neumann M, Notarnicola C. Review of machine learning approaches for biomass and soil moisture retrievals from remote sensing data. Remote Sensing. 2015;**7**(12):16398-16421

[50] Peng J, Loew A. Recent advances in soil moisture estimation from remote sensing. Water (Switzerland). 2017;**9**(7):1-5

[51] Amezketa E. Soil salinity assessment using directed soil sampling from a geophysical survey with electromagnetic technology: A case study. Spanish Journal of Agricultural Research. 2007;**5**(1):91-101

[52] Triki Fourati H, Bouaziz M, Benzina M, Bouaziz S. Detection of terrain indices related to soil salinity and mapping salt-affected soils using remote sensing and geostatistical techniques. Environmental Monitoring and Assessment. 2017;**189**(4):177

[53] Datta S, Taghvaeian S, Ochsner TE, Moriasi D, Gowda P, Steiner JL. Performance assessment of five different soil moisture sensors under irrigated field conditions in Oklahoma. Sensors (Switzerland). 2018;**18**(11):1-17

[54] Dursun M, Ozden S. A wireless application of drip irrigation automation supported by soil moisture sensors. Scientific Research and Essays. 2011;**6**(7):1573-1582

[55] Domínguez-Niño JM, Oliver-Manera J, Girona J, Casadesús J. Differential irrigation scheduling by an automated algorithm of water balance tuned by capacitance-type soil moisture sensors. Agricultural Water Management. 2020;**228**:105880

[56] Li M, Shamshiri RR, Schirrmann M, Weltzien C. Impact of camera viewing angle for estimating leaf parameters of wheat plants from 3D Point Clouds. Agriculture. 2021;**11**:6

[57] Soulis KX, Elmaloglou S, Dercas N. Investigating the effects of soil moisture sensors positioning and accuracy on soil moisture based drip irrigation scheduling systems. Agricultural Water Management. 2015;**148**:258-268

[58] El Marazky MSA, Mohammad FS, Al-Ghobari HM. Evaluation of soil moisture sensors under intelligent irrigation systems for economical crops in arid regions. American Journal of Agricultural and Biological Science. 2011;**6**(2):287-300

[59] Saleh M, Elhajj IH, Asmar D, Bashour I, Kidess S, Experimental evaluation of low-cost resistive soil moisture sensors. 2016 IEEE International Multidisciplinary

Conference on Engineering Technology (IMCET); 2016

[60] Jusoh M, Zuhariman N, Muttalib M. Efficacy of Arduino based low-cost Resistive Sensor in evaluating Soil Moisture from different Soil types collected in the Kelantan-Terengganu plain of Malaysia. IOP Conference Series: Earth and Environmental Science; 2020

[61] Qualls RJ, Scott JM, Deoreo WB. Soil moisture sensors for urban landscape irrigation: Effectiveness and reliability. JAWRA Journal of the American Water Resources Association. 2001;**37**(3):547-559

[62] Vivoni ER, Gebremichael M, Watts CJ, Bindlish R, Jackson TJ. Comparison of ground-based and remotely-sensed surface soil moisture estimates over complex terrain during SMEX04. Remote Sensing of Environment. 2008;**112**(2):314-325

[63] Matula S, Bát'ková K, Legese WL. Laboratory performance of five selected soil moisture sensors applying factory and own calibration equations for two soil media of different bulk density and salinity levels. Sensors (Switzerland). 2016;**16**(11):1912

[64] Gonzalez-Teruel JD, Torres-Sanchez R, Blaya-Ros PJ, Toledo-Moreo AB, Jimenez-Buendia M, Soto-Valles F. Design and calibration of a low-cost SDI-12 soil moisture sensor. Sensors (Basel). 2019;**19**(3):491

[65] Radi M, Ngadisih M, Nuha M, Alfinur RF. Calibration of Capacitive Soil Moisture Sensor (SKU:SEN0193). 2018. pp. 1-6. DOI: 10.1109/ICSTC.2018.8528624

[66] McKim HL, Pangburn T, Walsh JE, LaPotin PJ. Remote sensing of soil moisture. Advances in Space Research. 1987;7(11):139-145

[67] Adla S, Rai NK, Karumanchi SH, Tripathi S, Disse M, Pande S. Laboratory calibration and performance evaluation of low-cost capacitive and very low-cost resistive soil moisture sensors. Sensors (Switzerland). 2020;**20**(2):363

[68] Placidi P, Gasperini L, Grassi A, Cecconi M, Scorzoni A. Characterization of low-cost capacitive soil moisture sensors for IoT networks. Sensors (Switzerland). 2020;**20**(12):1-14

[69] Corwin DL. Soil Salinity Measurement. New York, NY, USA: Marcel Dekker; 2003. pp. 852-860. DOI: 10.1081/E-EWS120010191

[70] Heil K, Schmidhalter U. The application of EM38: Determination of soil parameters, selection of soil sampling points and use in agriculture and archaeology. Sensors (Switzerland). 2017;**17**(11):2540

[71] Clark PJ, Lamb DW, Bradbury R, Frazier P. The effect and mitigation of vine trellising on EM38 soil conductivity measurements. Proceedings of the Symposium on the Application of Geophyics to Engineering and Environmental Problems, SAGEEP. 2007; 800-811

[72] Slavich PG, Read BJ, Cullis BR. Yield response of barley germplasm to field variation in salinity quantified using the EM-38. Australian Journal of Experimental Agriculture. 1990;**30**(4):551-556

[73] Guo Y, Zhou Y, Zhou L. Using proximal sensor data for soil salinity management and mapping. Journal of Integrative Agriculture. 2019;**18**(2):340-349

[74] Hammam AA, Mohamed ES. Mapping soil salinity in the East Nile

Delta using several methodological approaches of salinity assessment. Egyptian Journal of Remote Sensing and Space Science. 2020;**23**(2):125-131

[75] Ding J, Yu D. Monitoring and evaluating spatial variability of soil salinity in dry and wet seasons in the Werigan-Kuqa Oasis, China, using remote sensing and electromagnetic induction instruments. Geoderma. 2014;**235-236**:316-322

[76] Guo Y, Huang J, Shi Z, Li H. Mapping spatial variability of soil salinity in a coastal paddy field based on electromagnetic sensors. PLoS ONE. 2015;**10**(5):1-12

[77] Bennett DL, George RJ. Using the EM38 to measure the effect of soil salinity on Eucalyptus globulus in south-western Australia. Agricultural Water Management. 1995;**27**(1):69-85

[78] Heil K, Schmidhalter U. Theory and guidelines for the application of the geophysical sensor em38. Sensors (Switzerland). 2019;**19**(19):4293

[79] Li Z, Wang N, Franzen A, Taher P, Godsey C, Zhang H, et al. Practical deployment of an in-field soil property wireless sensor network. Computer Standards and Interfaces. 2014;**36**(2):278-287

[80] Łukowska A, Tomaszuk P, Dzierżek K, Soil sampling mobile platform for Agriculture 4.0. 2019 20th International Carpathian Control Conference (ICCC); 2019

[81] Viscarra Rossel R, Adamchuk V. Proximal Soil Sensing2013. pp. 99-118. DOI: 10.13140/2.1.2799.1847

[82] Vaeljaots E, Lehiste H, Kiik M, Leemet T. Soil sampling automation case-study using unmanned ground vehicle. 2018. http://dx.doi.org/10.22616/ERDev2018.17.N503

[83] Bhanarkar MK, Korake PM. Soil salinity and moisture measurement system for grapes field by wireless sensor network. Cogent Engineering. 2016;**3**(1):1164021

[84] Wang Q, Terzis A, Szalay A. A novel soil measuring wireless sensor network. 2010 IEEE International Instrumentation and Measurement Technology Conference, I2MTC 2010—Proceedings. 2010;412-85415

[85] Shylaja SN, Veena MB. Real-time monitoring of soil nutrient analysis using WSN. 2017 International Conference on Energy, Communication, Data Analytics and Soft Computing, ICECDS. 2017:3059-3062

[86] Sui R, Baggard J. Wireless sensor network for monitoring soil moisture and weather conditions. Applied Engineering in Agriculture. 2015;**31**(2):193-200

[87] Zhang M, Li M, Wang W, Liu C, Gao H. Temporal and spatial variability of soil moisture based on WSN. Mathematical and Computer Modelling. 2013;**58**(3-4):826-833

[88] Balaji Bhanu B, Hussain MA, Ande P. Monitoring of soil parameters for effective irrigation using Wireless Sensor Networks. 6th International Conference on Advanced Computing, ICoAC. 2014:211-215

[89] Majone B, Viani F, Filippi E, Bellin A, Massa A, Toller G, et al. Wireless sensor network deployment for monitoring soil moisture dynamics at the field scale. Procedia Environmental Sciences. 2013;**19**:426-435

[90] Na A, Isaac W, Varshney S, Khan E. An IoT based system for remote monitoring of soil characteristics. 2016

International Conference on Information Technology, InCITe 2016—The Next Generation IT Summit on the Theme—Internet of Things: Connect your Worlds. 2017;316-320

[91] Li D, Gong J, Shan J. Geospatial technology for earth observation. Geospatial Technology for Earth Observation. 2009;**1**:1-556

[92] Inoue Y. Satellite- and drone-based remote sensing of crops and soils for smart farming–a review. Soil Science and Plant Nutrition. 2020;**66**:798-810

[93] Shamshiri RR, Bojic I, van Henten E, Balasundram SK, Dworak V, Sultan M, et al. Model-based evaluation of greenhouse microclimate using IoT-Sensor data fusion for energy efficient crop production. Journal of Cleaner Production. 2020;**263**:121-303

[94] Ge X, Wang J, Ding J, Cao X, Zhang Z, Liu J, et al. Combining UAV-based hyperspectral imagery and machine learning algorithms for soil moisture content monitoring. PeerJ. 2019;7:e6926

[95] Hu J, Peng J, Zhou Y, Xu D, Zhao R, Jiang Q, et al. Quantitative estimation of soil salinity using UAV-borne hyperspectral and satellite multispectral images. Remote Sensing. 2019;**11**(7)

[96] Luo W, Xu X, Liu W, Liu M, Li Z, Peng T, et al. UAV based soil moisture remote sensing in a karst mountainous catchment. Catena. 2019;**174**:478-489

[97] Zhang L, Zhang H, Niu Y, Han W. Mapping maizewater stress based on UAV multispectral remote sensing. Remote Sensing. 2019;**11**(6):1-24

[98] Zhang Y, Chen D, Wang S, Tian L. A promising trend for field information collection: An air-ground multi-sensor monitoring system. Information Processing in Agriculture. 2018;**5**(2):224-233

[99] Deng L, Mao Z, Li X, Hu Z, Duan F, Yan Y. UAV-based multispectral remote sensing for precision agriculture: A comparison between different cameras. ISPRS Journal of Photogrammetry and Remote Sensing. 2018;**146**(September):124-136

[100] Shamshiri RR, Hameed IA, Balasundram SK, Ahmad D, Weltzien C, Yamin M. Fundamental research on unmanned aerial vehicles to support precision agriculture in oil palm plantations. Agricultural Robots-Fundamentals and Application. 2018:91-116

[101] Allbed A, Kumar L. Soil salinity mapping and monitoring in arid and semi-arid regions using remote sensing technology: A review. Advances in Remote Sensing. 2013;**02**(04):373-385

[102] Sadeghi M, Babaeian E, Tuller M, Jones SB. The optical trapezoid model: A novel approach to remote sensing of soil moisture applied to Sentinel-2 and Landsat-8 observations. Remote Sensing of Environment. 2017;**198**:52-68

[103] Ahlmer AK, Cavalli M, Hansson K, Koutsouris AJ, Crema S, Kalantari Z. Soil moisture remote-sensing applications for identification of flood-prone areas along transport infrastructure. Environmental Earth Sciences. 2018;**77**(14):1-17

[104] Escorihuela MJ, Quintana-Seguí P. Comparison of remote sensing and simulated soil moisture datasets in Mediterranean landscapes. Remote Sensing of Environment. 2016;**180**:99-114

[105] Crosson WL, Laymon CA, Inguva R, Schamschula MP. Assimilating

remote sensing data in a surface flux-soil moisture model. Hydrological Processes. 2002;**16**(8):1645-1662

[106] Zhang D, Zhou G. Estimation of soil moisture from optical and thermal remote sensing: A review. Sensors (Switzerland). 2016;**16**(8):1308

[107] Fang B, Lakshmi V. Soil moisture at watershed scale: Remote sensing techniques. Journal of Hydrology. 2014;**516**:258-272

[108] Chakrabarti S, Judge J, Bongiovanni T, Rangarajan A, Ranka S. Disaggregation of remotely sensed soil moisture in heterogeneous landscapes using holistic structure-based models. IEEE Transactions on Geoscience and Remote Sensing. 2016;**54**(8):4629-4641

[109] Choi M, Jacobs JM. Soil moisture variability of root zone profiles within SMEX02 remote sensing footprints. Advances in Water Resources. 2007;**30**(4):883-896

[110] Yan H, Moradkhani H. Combined assimilation of streamflow and satellite soil moisture with the particle filter and geostatistical modeling. Advances in Water Resources. 2016;**94**:364-378

[111] Milzow C, Krogh PE, Bauer-Gottwein P. Combining satellite radar altimetry, SAR surface soil moisture and GRACE total storage changes for hydrological model calibration in a large poorly gauged catchment. Hydrology and Earth System Sciences. 2011;**15**(6):1729-1743

[112] Leng P, Li ZL, Duan SB, Gao MF, Huo HY. A practical approach for deriving all-weather soil moisture content using combined satellite and meteorological data. ISPRS Journal of Photogrammetry and Remote Sensing. 2017;**131**:40-51

[113] Er-Raki S, Chehbouni A, Duchemin B. Combining satellite remote sensing data with the FAO-56 dual approach for water use mapping in irrigated wheat fields of a semi-arid region. Remote Sensing. 2010;**2**(1):375-387

[114] Rahmani A, Golian S, Brocca L. Multiyear monitoring of soil moisture over Iran through satellite and reanalysis soil moisture products. International Journal of Applied Earth Observation and Geoinformation. 2016;**48**:85-95

[115] Wang L, Qu JJ. NMDI: A normalized multi-band drought index for monitoring soil and vegetation moisture with satellite remote sensing. Geophysical Research Letters. 2007;**34**(20):405

[116] Martínez-Fernández J, González-Zamora A, Sánchez N, Gumuzzio A, Herrero-Jiménez CM. Satellite soil moisture for agricultural drought monitoring: Assessment of the SMOS derived Soil Water Deficit Index. Remote Sensing of Environment. 2016;**177**:277-286

[117] Park S, Im J, Park S, Rhee J. Drought monitoring using high resolution soil moisture through multi-sensor satellite data fusion over the Korean peninsula. Agricultural and Forest Meteorology. 2017;**237-238**:257-269

[118] Kayode OT, Aizebeokhai AP, Odukoya AM. Soil salinity and its implications on sustainable agriculture in Southern and Northcentral States of Nigeria. IOP Conference Series: Earth and Environmental Science. 2021

[119] Wu W, Mhaimeed AS, Al-Shafie WM, Ziadat F, Dhehibi B, Nangia V, et al. Mapping soil salinity changes using remote sensing in Central Iraq. Geoderma Regional. 2014;**2-3**(C):21-31

[120] Thiam S, Villamor GB, Faye LC, Sène JHB, Diwediga B, Kyei-Baffour N.

Monitoring land use and soil salinity changes in coastal landscape: A case study from Senegal. Environmental Monitoring and Assessment. 2021;**193**(5):259

[121] Bouaziz M, Matschullat J, Gloaguen R. Improved remote sensing detection of soil salinity from a semi-arid climate in Northeast Brazil. Comptes Rendus—Geoscience. 2011;**343**(11-12):795-803

[122] Metternicht GI, Zinck JA. Remote sensing of soil salinity: Potentials and constraints. Remote Sensing of Environment. 2003;**85**(1):1-20

[123] Jiang H, Rusuli Y, Amuti T, He Q. Quantitative assessment of soil salinity using multi-source remote sensing data based on the support vector machine and artificial neural network. International Journal of Remote Sensing. 2019;**40**(1):284-306

[124] Abbas A, Khan S, Hussain N, Hanjra MA, Akbar S. Characterizing soil salinity in irrigated agriculture using a remote sensing approach. Physics and Chemistry of the Earth. 2013;**55-57**:43-52

[125] Guo Y, Shi Z, Zhou L. Integrating remote sensing and proximal sensors for the detection of soil moisture and salinity variability in coastal areas. Journal of Integrative Agriculture. 2013;**12**(4):723-731

[126] Casterad MA, Herrero J, Betrán JA, Ritchie G. Sensor-based assessment of soil salinity during the first years of transition from flood to sprinkler irrigation. Sensors (Switzerland). 2018;**18**(2):616

[127] Corwin DL. Climate change impacts on soil salinity in agricultural areas. European Journal of Soil Science. 2021;**72**(2):842-862

[128] Douaoui AEK, Nicolas H, Walter C. Detecting salinity hazards within a semiarid context by means of combining soil and remote-sensing data. Geoderma. 2006;**134**(1-2):217-230

[129] Eldeiry AA, Garcia LA. Detecting soil salinity in Alfalfa fields using spatial modeling and remote sensing. Soil Science Society of America Journal. 2008;**72**(1):201-211

[130] Fan X, Liu Y, Tao J, Weng Y. Soil salinity retrieval from advanced multi-spectral sensor with partial least square regression. Remote Sensing. 2015;7(1):488-511

[131] Yahiaoui I, Douaoui A, Zhang Q, Ziane A. Soil salinity prediction in the Lower Cheliff plain (Algeria) based on remote sensing and topographic feature analysis. Journal of Arid Land. 2015;7(6):794-805

[132] Allbed A, Kumar L, Sinha P. Mapping and modelling spatial variation in soil salinity in the Al Hassa Oasis based on remote sensing indicators and regression techniques. Remote Sensing. 2014;**6**(2):1137-1157

[133] Khanal S, Fulton J, Shearer S. An overview of current and potential applications of thermal remote sensing in precision agriculture. Computers and Electronics in Agriculture. 2017;**139**:22-32

[134] Venkatesan R, Tamilvanan A. A sustainable agricultural system using IoT. Proceedings of the 2017 IEEE International Conference on Communication and Signal Processing, ICCSP. 2018

[135] Shamshiri RR, Hameed IA, Karkee M, Weltzien C. Robotic Harvesting of Fruiting Vegetables: A Simulation Approach in V-REP, ROS and MATLAB.

Automation in Agriculture—Securing Food Supplies for Future Generations. 2018

[136] Loganathan GB, Mohan E, Kumar RS. IoT based water and soil quality monitoring system. International Journal of Mechanical Engineering and Technology. 2019;**10**(2):537-541

[137] Bhatnagar V, Chandra R. IoT-Based Soil Health Monitoring and Recommendation System. In: Pattnaik P, Kumar R, Pal S, editors. Internet of Things and Analytics for Agriculture. Singapore: Springer; 2020

[138] Raut R, Varma H, Mulla C, Pawar VR. Soil monitoring, fertigation, and irrigation system using IoT for agricultural application. Lecture Notes in Networks and Systems. 2018;**19**:67-73

[139] Siderius C, Biemans H, Conway D, Immerzeel W, Jaegermeyr J, Ahmad B, et al. Financial feasibility of water conservation in agriculture. Earth's Future. 2021;**9**(3):1-21

[140] Yamin M, Wan Ismail WI, Mohd Kassim MS, Abd Aziz S, Akbar F, Shamshiri R, et al. Modification of colorimetric method based digital soil test kit for determination of macronutrients in oil palm plantation. International Journal of Agricultural and Biological Engineering. 2020;**13**:188-197

[141] Maraveas C, Piromalis D, Arvanitis KG, Bartzanas T, Loukatos D. Applications of IoT for optimized greenhouse environment and resources management. Computers and Electronics in Agriculture. 2022;**198**:106993

[142] Shamshiri RR, Mahadi MR, Thorp KR, Ismail WIW, Ahmad D, Man HC. Adaptive management framework for evaluating and adjusting microclimate parameters in tropical greenhouse crop production systems. Plant Engineering. 2017:167-191

[143] Divya Vani P, Raghavendra Rao K. Measurement and monitoring of soil moisture using Cloud IoT and android system. Indian Journal of Science and Technology. 2016;**9**(31):1-8. DOI: 10.17485/ijst/2016/v9i31/95340

[144] Payero JO, Marshall MW, Farmaha BS, Davis RH, Nafchi AM. Development and application of cell-phone-based internet of things (IoT) systems for soil moisture monitoring. Agricultural Sciences. 2021;**12**(05): 549-564

[145] Jingwei W, Vincent B, Jinzhong Y, Bouarfa S, Vidal A. Remote sensing monitoring of changes in soil salinity: A case study in inner Mongolia. China. Sensors. 2008;**8**(11):7035-7049

[146] Shrestha RP. Relating soil electrical conductivity to remote sensing and other soil properties for assessing soil salinity in northeast Thailand. Land Degradation and Development. 2006;**17**(6):677-689

[147] Daliakopoulos IN, Tsanis IK, Koutroulis A, Kourgialas NN, Varouchakis AE, Karatzas GP, et al. The threat of soil salinity: A European scale review. Science of the Total Environment. 2016;**573**:727-739

[148] Babaeian E, Sadeghi M, Jones SB, Montzka C, Vereecken H, Tuller M. Ground, proximal, and satellite remote sensing of soil moisture. Reviews of Geophysics. 2019;**57**(2):530-616

[149] Petropoulos GP, Ireland G, Barrett B. Surface soil moisture retrievals from remote sensing: Current status, products & future trends.

Physics and Chemistry of the Earth. 2015;**83-84**:36-56

[150] Adamchuk VI, Hummel JW, Morgan MT, Upadhyaya SK. On-the-go soil sensors for precision agriculture. Computers and Electronics in Agriculture. 2004;**44**(1):71-91

[151] Dwevedi A, Kumar P, Kumar P, Kumar Y, Sharma YK, Kayastha AM. Soil sensors: Detailed insight into research updates, significance, and future prospects. In: New Pesticides and Soil Sensors. Elsevier Inc; 2017. pp. 561-594

[152] Athani S, Tejeshwar CH, Patil MM, Patil P, Kulkarni R. Soil moisture monitoring using IoT enabled arduino sensors with neural networks for improving soil management for farmers and predict seasonal rainfall for planning future harvest in North Karnataka-India. Proceedings of the International Conference on IoT in Social, Mobile, Analytics and Cloud, I-SMAC 2017. 43-48

[153] Rad AK, Shamshiri RR, Azarm H, Balasundram SK, Sultan M. Effects of the COVID-19 pandemic on food security and agriculture in Iran: A survey. Sustainability. 2021;**13**(18):10103

[154] Rad AK, Zarei M, Pourghasemi HR, Tiefenbacher JP. The COVID-19 crisis and its consequences for global warming and climate change. Computers in Earth and Environmental Sciences; 2022. pp. 377-385